Wilson Hamilton

The New Empire and Her Representative Men

Or, the Pacific coast, its farms, mines, vines, wines, orchards, and interests; its productions, industries and commerce, with interesting biographies and modes of travel

Wilson Hamilton

The New Empire and Her Representative Men
Or, the Pacific coast, its farms, mines, vines, wines, orchards, and interests; its productions, industries and commerce, with interesting biographies and modes of travel

ISBN/EAN: 9783337244613

Printed in Europe, USA, Canada, Australia, Japan

Cover: Foto ©ninafisch / pixelio.de

More available books at **www.hansebooks.com**

THE
NEW EMPIRE

AND HER

REPRESENTATIVE MEN;

OR

THE PACIFIC COAST,

ITS FARMS, MINES, VINES, WINES, ORCHARDS, AND INTERESTS; ITS PRODUCTIONS, INDUSTRIES AND COMMERCE,

WITH

INTERESTING BIOGRAPHIES

AND

MODES OF TRAVEL.

BY

WILSON HAMILTON.

OAKLAND, CAL.:
PACIFIC PRESS PUBLISHING HOUSE.
1886.

Entered, according to Act of Congress, in the Year 1886, by

P. M. Hamilton,

In the office of the Librarian of Congress, at Washington, D. C.

ALL RIGHTS RESERVED.

THE PACIFIC PRESS
PRINTERS, ELECTROTYPERS AND BINDERS,
OAKLAND AND SAN FRANCISCO.

TO THE

Representative Men of the New Empire,

THIS BOOK

— IS —

RESPECTFULLY DEDICATED

BY THE

AUTHOR.

CONTENTS.

Introductory	v
Topography	9
Productions of the New Empire	14
Vineyards and Orchards	28
Money, its Nature and Uses	36
Gold and Silver	39
Curbstone Brokers	41
Discovery of Gold	44
Dealing in Stocks	52
Other Minerals	55
Yosemite	60
School, Pulpit and Press	68
Bench and Bar	75
Monterey	77
Modes of Travel	80
World's Fair	83
California	85
Irrigation	93
Irrigation, Continued	98
Fresno County	106
Tulare County	109
Kern County	113
Los Angeles, San Bernardino, and San Diego	117
Merced County	121
The San Joaquin Valley and its Mountain Rim	125
Around the Sacramento Valley—The Mountain Counties of the North	128
The Sacramento Valley	132
California—A Resumé	138
Irrigation and Drainage	142
Biographies	166
John P. Jones	167
George Hearst	171
Leland Stanford	172
James G. Fair	182

INTRODUCTORY.

FOR the purposes of this work, all that scope of country west of the Rocky Mountains is embraced in the domain designated as the New Empire, but for a more definite and comprehensive description of its extent, coast line, political divisions and population, the following table is submitted:

POLITICAL DIVISIONS.	Area in Square Miles.	Population.	Inhabitants to Square Mile.	Length on Meridian, Miles.	Coast Length, Miles.	Full Shor Line, Miles.
California	185,360	864,686	5	655	735	1,007
Oregon	96,030	174,767	2	288	300	285
Nevada	110,700	62,265	½	440	....	....
Washington	69,180	75,120	1	245	245	1,738
Idaho	84,800	32,611	1⅓	480	....	....
Utah	84,970	143,906	2	345	....	....
Arizona	113,020	40,441	1⅓	480	....	....
Pacific Montana	22,000	8,000	1¼	310	....	....
Pacific Wyoming	22,000	4,000	¼	275	....	....
Pacific Colorado	40,000	10,000	¼	275	....	....
Pacific New Mexico	24,000	20,000	1	345	....	....
Alaska	577,390	30,000	....	1,190	1,470	20,000
British Columbia	310,000	20,000	....	306	560	8,181
Pacific Mexico	550,600	3,500,000	7	1,260	1,650	4,000
Pacific Central America	50,000	600,000	11	560	910	1,450
Total	2,312,450	5,585,796	2	...	6,170	36,751

This gives an area of 2,312,450 square miles, extending over a territory ranging from perpetual winter to eternal spring, so varied in its climate and productions as to yield almost every article requisite for the use of man, indeed, so broad and varied in its range, that, were the Pacific Coast shut off from the rest of the world by an impassable gulf or blockade, the population, enlarged to 100,000,000 or more, could still live in the enjoyment, by production and manufactories, of every necessary and luxury of life. Of this area 250,000 square miles may be considered valuable chiefly for the minerals which are found, of great richness and variety, in the river beds, gulches and mountain ranges; an equal amount is at present regarded as comparatively barren and valueless; 500,000 square miles is tillable soil, 300,000 excellent forest, and 1,000,000 good grazing land.

It is the design of the author to make "The New Empire and Her Representative Men" a popular book, replete with practical information on subjects of general interest relating to the region west of the Rocky Mountains, showing the rapid progress which has been made since the discovery of gold in 1848, in the creation and accumulation of wealth and in advanced civilization, and also the beneficent opportunities yet waiting to be appropriated by the intelligent and industrious immigrant. Thirty years ago this whole realm was a comparatively uninhabited wilderness; to-day it is a thriving, prosperous commonwealth, peopled by an enterprising, industrious population, successfully engaged in all the various avocations of life.

No State or country can be found, from the rising to the setting of the sun, where local institutions are more diversified, where they have grown up more rapidly and been established on a firmer or more enduring basis than here.

Mining was the original, and, for a long time, the chief business of the coast; but, as mining developed, broadened its area and influence, it created new demands, demands for the brightest intellects in the professions, in the trades, and in commerce, for banking, mercantile and shipping houses, for steamship lines to traverse the ocean in all directions, and our bays and rivers at home, for transcontinental and local railroads, for manufacturing establishments, insurance companies, all the productions of the husbandman, the vineyardist and kindred industries.

And, whilst the people have been pushing, crowding, grasping and hurrying to get rich, they have steadily cultivated the higher qualities of manhood, morally, socially and politically; they have erected magnificent churches and contributed liberally to their support; built up and endowed institutions of learning, of the arts and sciences which are the pride and ornament of the coast.

As a rule they have learned to accept the ups and downs, the stern realities of life, philosophically, neither to be over elated as the toiler in poverty suddenly drops into the lap of luxury, or to be depressed when the tide of fortune turns and sweeps him down the stream, recognizing the fact that every man must fight the battle of life for himself, and that it is the part of good citizenship to do it independently and cheerfully.

In this regard our representative men have set an example worthy of emulation by devoting a share of their talents, influence and wealth to the good of their fellows. Bold and determined in business, vigorous and comprehensive in thought, generous and manly in spirit, they engage in the most stupendous undertakings, as though they were but every-day affairs, and then energetically push them to success and so accumulate colossal fortunes. They do not live for themselves alone, but recognize their duty and responsibility to the State and society, ready and willing of their own abundance to create new enterprises, if need be, or foster those of others, whether of a public or private nature, and for the unselfish purpose of doing good to the community in which they made their money, satisfied with the rewards that good deeds always bring. The Pacific Coast has many such men, and it is fortunate that she has, because it is to their commendable enterprise, broad-reaching views and business tact, that we owe the development of our resources and the prosperity we enjoy.

But there are non-representative men, those who represent nobody and nothing but themselves and their individual estates; misers, measly misers, narrow-minded money gluttons who acknowledge no duty or obligation to the State or society, who shut themselves up with their gold bags and turn their backs on every enterprise and every good work without regard to its merits or necessity. The good they do is when they die; the joy they bring is when relatives gather around the executor's table to receive their share of the hoarded treasure.

The writer was a pioneer; has seen all the phases, the bright and shady side, of California life; familiar with historic facts and events, he will treat all the subjects incident to "The New Empire and Her Representative Men," associated with *living issues*, faithfully, from personal knowledge and observation, properly representing the character of the country for health, wealth, natural resources, climate and scenery, and pointing out clearly and concisely the advantages and inducements offered to those seeking homes and fortunes on the friendly shores of the Pacific.

THE NEW EMPIRE

AND HER

REPRESENTATIVE MEN.

CHAPTER I.

TOPOGRAPHY.

THE topography of the Pacific Coast, and of California especially, greatly resembles that of Asia. The Sierra Nevada Range of mountains rises like a rampart, lofty, mysterious, snow-crowned, along the eastern line of the State, furnishing scenery as varied and as grand as the eye of man has seen. High up on their crest are the head-waters of great rivers, and there lakes nestle under the guardianship of the clouds. In one place, slashed by that God-wrought wonder, the Yosemite Valley; in others, fissured by profound cañons, their slopes are shaded by forests of pine and cedar, and their granite frames nurture the great Sequoia, the big trees, over which the world has marveled. Out of their sides burst hot and mineral springs, with high medicinal and curative properties, and vineyards are creeping up the terraced grade of their foot-hills. Set along them are the craters of volcanoes extinct, great scars of a fiery ulceration, that mark the long past period of upheaval. In other places lone peaks uncover their blear skulls to the storm and sunshine, far above the spurning and conquering foot of the explorer.

Between this range and the Coast Mountains, lie the two great valleys of California, the San Joaquin and Sacramento, traversed by the streams of the same names, which receive many a snow-fed confluent, and are wedded in the waters of San Francisco Bay. These valleys cover 64,000,000 of acres, and with proper conditions it is all tillable and capable of high farming. Sheltered by the lofty mountains, they are the home of the vine and olive, and of all the semi-tropical fruits. In the spring-time they are closely carpeted with wild flowers of many colors, which reach beyond the vision in solid masses of gay tint. Soon human industry will cover these with vineyards, and wrest from them a harvest of delicate and necessary food that will make American markets independent of the raisins of Valencia, the oranges of Messina, and the oils of Lucca.

Following the western rim of these valleys, the Coast Range rises and shelters charming vales and glens highly cultivated, and sustaining a prosperous population. In this Coast Range are the dairy pastures of the State, and, as they are developed, their herbage will send out cheese that rivals that made in the vales of Cheddar, equal to the Neufchatel and *fromage de Brie*.

These valleys were all once the feeding ground of countless herds of cattle and droves of horses. The latter would so increase that long before American occupation they would be circled in a grand *battue* and stampeded over the cliffs into the Pacific. And to this day, along the coast is many a Golgotha covered with reefs of their bones. In the mountains on either side of these main valleys, are the world's richest mines of precious metals. Here are gold and silver and quicksilver, and the torrents of ages have washed gold into the beds of the streams, where it lies, a tempting prize, in many a natural sluice-box, caught in the rocky riffles invented by nature long before a "long tom" was devised by man.

On the Nevada side of the Sierra Nevada Range, the baldness of the mountains is compensated by the richness of the mineral deposits which they hide, while they bound, also, many a green valley, fairy lake and brawling stream, and on all sides rise so as to shut in the State by mighty ramparts that make it like a great dish; and all the streams that rise within its borders sink also within them. Nevada keeps her waters at home, and gives none to the riparian systems around her, and none to the full, yet thirsty sea.

The mountain system of Oregon in general resembles that of California, and is closely copied by Washington Territory.

The most remote of our possessions lies still beyond, and the mountains which, south of the Rio Grande we name the Cordilleras, the Andes, and the Sierra Madre, in California the Sierra Nevada, link Alaska to the tropics; albeit, here they guard fiords as grand as those of Norway, and down their cañons creep glaciers to which science and curiosity will make pilgrimage.

Here we have the climate of Scotland in the latitude of Scandinavia. The waters are crowded with fish, and the rocks peopled with seal and sea otter, the noblest of fur-bearing amphibia, while the hills are richer in coal than all England, Scotland, and Belgium, and even superficial search has revealed gold mines where the ore is stripped like a limestone ledge, quarried with a crowbar, and dumped into the mills that have tide-water on their outer wall. Here, too, are forests as dense and trees as grand as those in whose shadows human fancy wrought out the images of Thor and Woden, as children see pictures in the fire; and the whole, fish, fur, timber, coal, and gold, offer virgin resources awaiting to be made productive by wedding them to human skill and energy.

The geology of all this region of mountains, and foot-hills, and valleys, belongs, as to its rocks, to the plutonic, upper secondary, tertiary, and volcanic formations. Here nature set up her anvil, and from the fires of her forge welded the spine and ribs of a structure that rose over against the sea and put an everlasting bound to its waters. Then, upon the sublime heights, came snows and floods, and by erosion, corrugated the mountain-sides with gorge and cañon, and in the process, disintegrated granite and the metamorphic rocks to make the soils and sands of the valley. Volcanic fires and forces spouted lava to run like rivers searching for the sea, and as sun and shower comminuted and dissolved it, to prepare food for the vine and olive.

The botany and zoology of all this area are set with features not held in common with any other part of the Union. Even the deer differs from its cousin east of the Rocky Mountains, and the wild goat and mountain sheep have no familiar representative elsewhere. Here is the home of the grizzly, the monarch of bears, and of the mountain lion, that scourge of the sheep-fold, and wily enemy of man. Even the robin and jay wear here a different plumage, and the very lark salutes the cheery morning with a novel note.

The oak, elm, and willow are peculiar, and so through the whole range of deep vegetation. The big trees are the last of the giant *autocthons* that were before the forests of spruce and pine and cedar had been nurtured in their shade. Here the bay tree distills its spicy odors, the madrona spreads its tawny arms, and the manzanita softens the landscape with its dark red bark and foliage of steely green. The very flowers are diverse in their beauties, from the voluptuous lily of Mariposa to the golden poppy and blue lupin, which carpet the plains in fabric of color ever changing, and always beautiful.

The scenery furnished by this variety and combination of mountain, plain and valley, tree and blossom, has no superior in the world. No wonder that here the brush of Bierstadt caught its earliest inspiration, and did its noblest work. Here the sun shines more hours in the year than elsewhere; the climate conduces to the highest state of health, is the most equable, and permits the soul to know the presence of the encumbering body only through the sensations of pleasure and peace of which it is the medium.

Though a winterless land, this is not a climate that enervates, but, to the contrary, it seems to spur men to the keenest quest, to sharpen their faculties for industry, and in witness of this stand our railroad constructions, the most difficult in the world, our delta lands, redeemed as Holland conquered provinces from the ocean, and our deep mining, which overcame the most appalling subterranean problems, and taught a novel pathway to mineral treasures secreted in what may be termed the most intricate convolutions of the bowels of the earth.

"A HOME IN THE NEW EMPIRE."

CHAPTER II.

THE PRODUCTIONS OF THE NEW EMPIRE.

THE Pacific slope of the American Continent is a country of great natural resources and productions. Her mineral deposits include every variety in general use. Her forests contain every useful tree, even those nowhere else found, with wild fruits in profusion, and a charming variety of shrubs and flowers to adorn and beautify. Her waters abound in food fish and fur-bearing animals. Her mountains, forests and plains are alive with a wide range of game, fur and feathered, large and small. And wheat, oats, barley, rye, corn, peas and beans, with staple vegetables and fruit, grow abundantly all over the realm, while California excels in semi-tropical productions, and in certain localities even those indigenous to the tropics. Her scenery has a broad and interesting range from the pastoral and gentle to the wild, imposing and impressive. Her climate is full, clear, and bracing, invigorating and healthful; conducive to long life and happiness.

The nature and yield of the crops of any country depend on climate, soil and other conditions. On the Pacific slope it is the other conditions which strike the new-comer and even the old settler with astonishment. They are subtle, peculiar and cannot be explained on any general known law or principle; and as this is not an abstruse but rather a practical work, no theories will be indulged in, or effort made seeking an explanation, but the facts will be accepted as found and intelligently applied. The British poet laureate can take a worthless sheet of paper, and, by writing a poem on it, make it worth $65,000; that is genius. A millionaire can write a few words on a sheet of paper and make it worth $5,000,000; that is capital. The United States can take an ounce and a quarter of gold and stamp an eagle bird on it, and make it worth $20.00; that is money. The mechanic can take the material worth $5.00 and make it worth $100; that is skill. The merchant can take an article worth 25 cents and sell it for $1.00; that is business. These things are peculiar to the parties in interest, but they can all be explained.

Indigenous animals, trees, fruits and flowers are found here under conditions nowhere else to be met with. The cereals, fruits and blossoms transplanted here from their native or other soils excel in abundance, richness and flavor under conditions nowhere else found. This is peculiar to the Pacific Coast, but it cannot be explained, and hence explorers, travelers and tourists exclaim, " We never saw the like of this before;" and scientific climatologists confess themselves balked by things they cannot account for.

CEREALS.

In order to give the reader as clear and perfect an understanding as possible of the situation, the yield and value of the cereal crops, the writer has selected from a vast mass of matter gathered during the past year, a number of statements made by settlers residing in different sections of the country, and will give them in the exact language of the farmers who made them, because they are reasonably broad in their range, and are evidently the candid expressions of opinions by well-disposed men, who, from practical tests and experience, know of what they speak and can testify of what they have seen. " I came here," says the first farmer, " on April 3, 1877, and made a homestead settlement on this land, and have therefore been here eight years. To say that I am glad that I came would but mildly express my feelings on the subject. The first year I put in 15 acres of wheat, oats, and barley, and about 1 acre of potatoes. My wheat, sown on the sod, brought 25 bushels to the acre; barley 30, and oats 50 bushels to the acre. The potatoes, also planted on the sod, yielded 180 bushels to the acre. I think this is the best poor man's country in the United States, and the healthiest. It is far ahead of Texas. There is no man who can come into this country and fail to prosper and get a home, if he can work and is industrious. When I came I had nothing except an old wagon and four old mustang horses. I came overland. When I reached here I had only $75.00 left, and with this I bought my seed, plow, 15 harrow teeth, and groceries for 6 months, and I had $1.75 left. The first thing I did was to commence plowing. The next year I broke up about 12 acres more, making 27 acres in all, and sowed it to wheat, oats and barley. That year the old and new wheat ground averaged 35 bushels per acre. In the following year, I broke up a little more ground, but had only 3 acres in wheat, which averaged 57 bushels, and it was the best I ever raised. I threshed this out with the horses, and if I had had a threshing machine it would have averaged a great deal more. In regard to the general run of the crops here, I have raised about 35 of wheat, average, and barley 55 to 60. All the wheat I have raised has been spring wheat, sown from the 1st of May to as late as the 15th of June. It would not be safe to sow it later than the 15th of June, though it might be put in as late as the 15th of July and if the season happened to be of unusual length a crop of grain might be reaped; and, if not, a good crop of hay could be cut off. In subsequent years my crop averaged just about the same as the first year. In 1883 my wheat went 33 bushels to the acre; barley went about 60 and oats about 75. Last year I sold my produce right here on my place and did not haul a pound away. My wheat brought me $1.00 a bushel; barley 96 cents; oats 63 cents. Butter will average 25 to 50 cents per pound; chickens from $3.50 to $6.00 per dozen, and eggs 25 to 40 cents. This is a superior country for chickens. Hogs 3 to 5 cents on foot. Dressed pork 4 to 7 cents per pound; bacon, now selling at 16 to 18 cents, has been selling as high as 20. Stock cows with calves at their side are worth $12.50 per head; dairy cows with calves, $25.00 to $40.00 per head. I am well acquainted with all the country, and have kept posted as to the yield of crops, and from what I know it would not be far out of the way to place the average yield of produce, grain, etc., about as follows: Wheat, 35 bushels; oats, 75; barley, 50; corn, 40; rye, 20; beans, 30; potatoes, 300. To prove what I say, that this is the best poor man's land in America, I will figure up the result of my 8 years' labor on this farm: I have here 160 acres of land, which, in its present improved condition, with houses, barns, fences, etc., is worth $20.00 per acre, and I would not sell at that price—land of the same quality as millions of acres in this neighborhood belonging to the Government and the railroad. I have 11 head of horses, worth at least $50.00 each; I have 16 dairy cows, worth $35.00 each; 16 head of young stock on the range, yearlings and 2-year-olds, worth $160; a mower and reaper combined, hay rake, a sulky plow and a new Mitchell wagon; a harrow, cultivator, and all other tools and implements necessary for my work, which are worth $602. And I realized out of my last year's crop, clear of all expenses, $635. To sum up, I have:—

Land worth	$3,200
11 head horses	550
16 dairy cows	560
Cattle on range	160
Farming implements, etc.	602
Net profit on last year's crop	635
Total	$5,707

"To this aggregate should be added an allowance for food and clothing for myself and family during the eight years which, at a very low estimate, was not less than $400 per year, making for the whole time, $3,200. When you add this to the value of the land, stock, etc., you have a total of $8,907. Therefore, by a little figuring, you will see that I have earned about $1,100 per year. Now, in addition to this, I have bought 320 acres of railroad land at just a little more than the price of Government land, and in two years my 480 acres will probably be worth $50.00 per acre. To state the case roughly, ten, or at the most twelve years of healthful occupation and labor in one of the very best of climates, with only a couple of hundred dollars in cash and effects to start on, will have accumulated for me money and property to the amount of $25,000. And if you can show me any other country in which a man can do the like of this, I would just like to have you point it out to me."

The next farmer, residing in one of the Territories, says: "I have been on this land four years, and have raised four crops. The first year I broke up 15 acres, plowed the last week in May, sowed wheat, and threshed out by tramping of horses 45 bushels to the acre. I sowed about the 10th of June. I had also 20 acres of oats, which were as high as a man's head, and brought me 75 bushels to the acre. Another field of 30 acres was broken up that summer, and in the following season, April 1, sowed to wheat; and this time, by the use of the threshing machine, I saved a yield of 1,650 bushels, or 55 bushels per acre. It stood six feet high, and headed out larger than any wheat I ever saw. In the course of a few years there will be no more open stock range left in this region, for wheat growing will take up all the land. My crops have been very large every one of the seasons I have been here. I do not hesitate to say that the average wheat crop for last year was over 35 bushels to the acre. I know this sounds big, and people might be tempted to say I was willfully misrepresenting the facts, but a good many things happen in this world from time to time that are not put down in the books and that are new to the common experience."

It may be instructive to look for a moment at the profit that may be realized from a quarter-section, yielding in wheat the average indicated. The following estimate is made from data taken from sources of authority:—

EXPENSES.

Fall plowing 160 acres, at $2....	$320
Seed wheat, 1½ bushels per acre, at 45c.	108
Sowing and harrowing, 75c. per acre.	120
Cutting, binding, and shocking, $2 per acre.	320
Hauling, threshing, etc., $2.50 per acre.	400
Total expenses of crop, $7.92½ per acre.	$1,268

RECEIPTS.

3,600 bushels of wheat, being an average yield of 35 bushels per acre, worth 50c. per bushel.	$2,800
Deduct expenses of crop, $7.92½ per acre.	1,268
Receipts over expenses, $9.57½ per acre.	$1,532

Here is a profit per acre of nearly four times the price at which the land is offered for sale to-day, both by the Government and the railroad.

Another farmer, residing in the North, says: "I came from New York and settled here about 15 miles from a railroad station, and have 320 acres of land. All of my farm except 80 acres is bottom-land, watered by a creek and springs. The 80-acre tract is on the hill-sides, and was originally bunch grass land. I have raised 100 bushels of the best wheat I ever saw, on less than one acre, the field which you see in front of my house. I have sowed 30 acres of wheat every year since I came here, and the average for 7 years past on these 30 acres has been 45 bushels per acre. These 30 acres produced from 35 bushels on the poorest ground to 100 on the best, averaging as I have told you. There is plenty of just such land as this, where the soil is darker and is not made land by washings from the hills, but is black, rich land. Much of the choice Government land in the country is taken up by settlers, but there is plenty of land remaining which is for sale by the railroad. The best land is of the high rolling prairies, with clay subsoil, the top soil being from one to six and more feet deep. The value of clay subsoil is that it holds the rains and moisture longer. The deeper the body of clay, the better the wheat land, but this is owing only to the fact that the clay retains the moisture, not permitting it to sink out of reach of the surface soil. I guess that the presence of clay has nothing to do with the quality of the land, but is only a conservator of moisture. Hence it is that land having a clay subsoil may be of a poorer quality than other land not having it, and yet produce a larger crop. There is sufficient moisture in the atmosphere to produce a large yield without the help of rain. The seasons of 1872-73 and 1874, were the driest seasons since I came here, during which but a very small quantity of rain fell. In 1872, from March 15 to December 15, there was not more than three or four showers, and those very light, falling in July, and yet the crops those years were the heaviest I have raised here. It was in one of these dry seasons, 1873, that I raised over 100 bushels of wheat to the acre. The heads were exceedingly long and heavy; and, mind you, this wheat was not threshed with a machine, but tramped out in the old-fashioned way. I have a pretty fair idea of the character of the county.

"Besides wheat, we can raise crops of other kinds that would surprise people. This land yields 60, 70, and 80 bushels of oats to the acre; and I have never seen so poor a crop here as 30 bushels per acre. Corn does well on high lands

ANOTHER HOME IN THE NEW EMPIRE.

and on the tops of hills, but not in the bottoms. It would average say 40 bushels to the acre. The reason why it does not grow well in the low-lands is that there is too much frost there at night; but there is none on the hills. Peas are the largest crop you can put in the ground. I have raised over 50 bushels to the acre. As to hay, there has not been any sown to speak of. But the wild timothy grows profusely. I have cut 2½ tons to the acre of this wild hay from the bottom-lands. The temperature in this part of the county is more even than that of any country I have ever lived in. The winters here are tropical compared to those of Illinois. Because of the mildness of the climate and the cheap and good grazing, this is the best sheep and cattle country that I have ever been in. Finally, it is not, in my judgment, extravagant to say that the average yield of wheat raised last year was fully 35 bushels to the acre. I would not put it below this figure, and some people believe it ran above. It is going to be very profitable to raise wheat in this country.

"I do not see why a poor farmer, having a quarter-section of land here, could not, by ordinary prudence in management, not only be independent, but make good headway in a few years toward comparative wealth. There is certainly a great future for this region, and for the men who are lucky enough to secure farms for themselves before these lands rise much in value, as they must soon do."

An able and scientific writer says: "It is a known fact that the most productive and enduring wheat lands of our continent, lie west of the Rocky Mountains. They have the largest proportions of the potash and phosphates which nourish the cereals. It has been stated by a well-known geologist that, during the six distinctly noted volcanic overflows, the ashes, which were carried largely by the prevailing winds eastward into the bays and lakes which formerly occupied the great interior basin, mingled with other sediment to form the deep deposits which now constitute the soils of those valleys and high prairie lands. It is easy to infer that the excess of alkali in spots, results from the drainage of this substance from the hills. Every year the crops seem to increase in value and amount. The hills and dry sage-brush plains have rewarded the cultivator. It is known that every acre touched by water becomes luxuriant with cereals and fruits.

"It is known that an ocean of aerial moisture floats over these regions from the vast western ocean. It needs only a cooler to deposit the dews. Every field or blade of grass or grain acts as a cooler.

"The fields of winter grain, started by early rains or melting snows, provide the vegetation, which in summer deposits enough of this aerial moisture to perfect their growth until the harvest. The deep-plowing loosens the soil so as to absorb the air loaded with moisture, which grows cool enough to leave its moisture about the roots of the plant. Thus the lands that have for ages abounded in the bunch grass, which is now wasting away before the increase of flocks and herds, can be restored by the plow, and the choice cereals, wheat, oats, barley, and corn, with orchards about every farm-house."

Hundreds of other experiences on the Pacific slope could be given, but these are enough; they reflect the experience of thousands. Are they not satisfactory? How could they be better? Compare them with the average yield of wheat in the principal Atlantic States for 1880. Maine's average yield, per acre, 14 bushels; New Hampshire, 14; Vermont, 17; Massachusetts, 22; Connecticut, 13; New York, 19; New Jersey, 15; Pennsylvania, 15; Delaware, 13; Maryland, 13; Virginia, 7.2; North Carolina, 6.5; South Carolina, 5.5; Georgia, 7; Alabama, 7.3; Mississippi, 6.8; Texas, 16; Arkansas, 6; Tennessee, 5; West Virginia, 11.5; Kentucky, 9.3; Ohio, 18; Michigan, 18.3; Indiana, 16; Illinois, 13.6; Wisconsin, 12.4; Minnesota, 12; Iowa, 9.4; Missouri, 11; Kansas 16.3; Nebraska, 13.1.

It will be seen by these figures that the yield of wheat in the most favored of the Atlantic States falls much below the yield here, and even then it depends on the use of costly manures. As the production depends on conditions, so the value of wheat when garnered ready for market depends on conditions, facilities for transportation, etc. In Nebraska, where the transportation is arbitrary and limited, the average price per bushel is 42 cents; in Kansas, 45; in Dakota, 46; in Minnesota, 50; in Iowa, 55; and in Missouri 62. Here it will average 63 cents, which is a third more than in Nebraska, and higher than any of the western Atlantic States.

In the Pacific States and Territories there are 500,000 square miles, or 320,000,000 acres of tillable land. Setting aside one-half, or 160,000,000 acres of this for oats, rye, corn, barley, and other farm products, and presuming that the remaining 160,000,000 acres were devoted to raising wheat at the reasonable average of 20 bushels to the acre, we would produce 3,200,000,000 bushels, worth, at 63 cents a bushel, $2,016,000,000; and to move it would require a train of cars 35,058 miles in length, long enough to girdle the earth; or it would load the entire merchant marine of the world, sailing and steam vessels, a dozen times over.

In contrast to this we are reminded that the average acreage given to wheat raising in all England, from 1867 to 1870, was 3,836,890 acres; but there has been a gradual falling off every year since then, and last year only 2,553,092 acres were apportioned to wheat, an area about equal to one of our little valleys skirting the Columbia or nestling among the foot-hills up in Tulare.

CATTLE.—The live stock business in the United States has recently increased to enormous proportions. It is estimated that there is now invested in cattle alone, $1,106,715,-703. When to this is added all the industries and interests dependent on the cattle trade, these figures would be immensely increased.

In 1880, the estimated value of the meat, hides, and other proceeds of animals slaughtered in the United States, is fixed at $800,000,000.

West of the Mississippi there are 22,000,000 head of cattle, with an estimated value of $533,650,875. The business is chiefly carried on by cattle companies. One association alone owns 1,000,000 head of cattle, 1,000,000 head of sheep, and 500,000 head of horses, among which are to be found some of the best-bred animals in the world. The company also owns and controls 8,500,000 acres of land.

The total value of stock and land is set down at $68,250,000, and the company employs 2,000 men as herders. Another association has 450,000 head of cattle, 50,000 head of horses, 30,000 sheep, and 4,500,000 acres of land, with a total valuation of $21,500,000. Another company owns 800,000 head of cattle, 250,000 head of horses, and as many sheep, with a grazing area covering 15,000,000 acres of land. And there are many other companies owning immense herds, and controlling millions of dollars of capital.

On the Pacific slope there are 640,000,000 acres of excellent grazing land. The area required to pasture a million head of cattle depends entirely upon the quality of the grass and water. Here where our indigenous grasses are unusually nutritious and sweet, and the water surpassingly pure, and where the herds have abundant shade in the summer and shelter in the winter, a much greater percentage of cattle can be carried, and the losses much less than in localities where such advantages are not enjoyed. Utilizing the 640,000,000 acres of grazing land in the New Empire, allowing 5 acres to the animal, or even 20 acres, and our herds would number 152,-000,000 head, and at an all round average of $25.00 per head, their value would be $3,800,000,000, an excess of the total amount now invested within the American domain.

LUMBER.—There are 300,000 square miles of forest in the New Empire, a considerable portion of which is very superior. The trees not only stand thickly on the ground, but they are of immense size and height, a single tree frequently making five, ten, and fifteen thousand feet of clear lumber. By a little calculation the reader will be able to form some estimate of the amount of lumber and wealth contained in this 192,000,000 acres of timber.

A number of large saw-mills are at work in various places convenient to tide water, mortising into the forest and sending out lumber to the markets of the world. As an illustration of the magnitude of these milling enterprises, take the Puget Mill Company. It has a capital of $2,000,000, and has mills at Port Gamble, Port Ludlow and Utsalady, whose output in 1884 was 57,000,000 feet of lumber, worth

$741,000; shingles, 2,700,000, valued at $8,000; lats, 18,-000,000, valued at $36,000; pickets, 225,000, valued at $2,700; wool slats, 60,000, valued at $360; and 3,000 piles valued at $11,500, making a total value of $800,410. The Hanson Mill Company has a capital of $1,000,000; output last year, 33,000, value, $426,000; spars, 600, value, $12,500; laths, 6,500,000, value, $16,250; pickets, 350,000, value, $2,800; wool slats, 150, value, $900; total value, $458,450, with the addition of $40,000 as the product of the planing mill, swelling the total to $498,450. The Hanson Mill has since been enlarged, so that its output will now about equal the Puget Mill Company. These illustrations are sufficient. I do not know the total number of saw-mills on the coast, but there are a great many.

Lumber, like wheat, is a staple, cash commodity, and all these large mills, steadily manufacturing to supply foreign demand, bring back in return millions of money to be distributed annually among the people of the Pacific Coast.

The Census Department at Washington, in its forestry bulletins, announced that in both the upper and lower peninsulas of Michigan there remained of standing white pine timber, suitable for market, but 35,000,000,000 feet, board measure, and that in the census year of 1880 there had been cut in the State 4,396,211,000 feet, requiring only 8 years at this rate to exhaust the supply; that in Wisconsin there were standing 41,000,000,000 feet, with a cut of about 3,000,-000,000 feet for that year, leaving a supply that would last but 14 years; that in Minnesota there were remaining 8,170,-000,000 feet, and that 541,000,000 feet were cut in the census year, leaving a supply for 15 years, and that at this rate the supply of white pine lumber would be exhausted in these 3 States in the brief period of about 12 years, the question of the future supply of this most valuable timber became serious to the building world. The late James Little, of Montreal, in 1882, said of the supply of white pine in Canada that he had consulted with the best authorities and was persuaded that, at the rate of cutting then going on, the whole supply of the provinces of Quebec, Ontario, New

Brunswick and Nova Scotia would be used up in about 10 years. According to these estimates, then, the supply of white pine on the Atlantic slope will soon be exhausted and the mechanic arts will have to look to other fields for their supply of wood and timber, and that supply will be furnished by the forest of the Pacific Coast.

Red and white fir, pine, spruce, cedar, cottonwood, balm, oak, alder, ash, and maple are generally found in the principal coast forests, but California is the home of the redwood. The total production and sales of redwood in the State for 1885, from the several counties, were as follows:—

Del Norte County, feet,	4,050,000
Humboldt,	82,300,000
Mendocino,	74,050,000
Sonoma,	4,400,000
Santa Cruz,	40,000,000
Total feet,	204,800,000

Of the sales in Santa Cruz County, it is estimated that 35,000,000 feet were taken for consumption in that county, while the other 5,000,000 feet were shipped to points south. The other counties sent to the Bay of San Francisco 113,000,000, besides 32,150,000 feet to Southern California, and 19,650,000 feet were sold for consumption in the counties where cut. The shipments of California redwood to foreign ports in 1885 were as follows:—

Mexico, Central and South America, feet,	950,000
Islands of the Pacific,	2,650,000
Australia,	5,950,000
Europe,	650,000
Total feet,	10,200,000

The pine sales from redwood mills for the year aggregate 17,800,000 feet. This makes the total sales for the year 232,800,000 feet. The quantity of redwood on hand January 1, 1886, was 41,350,000 feet, against 34,040,000 feet on the 1st of January, 1885, but the demand is increasing as its uses multiply, and its value and beauty are demonstrated. It is beginning to be largely used in the manufacture of the most elegant patterns of furniture, and the roots and buhl of the redwood tree are now coming into extensive use for veneer-

CAMPING IN THE FORESTS OF THE NEW EMPIRE.

ing purposes, furnishing a more beautiful veneer than mahogany, rosewood or walnut. A machine has lately been invented and patented for slicing the roots and buhls for veneering purposes, and a piece of root no larger than a man may carry on his shoulder brings $20.00 in Europe.

IRON.—Iron, coal and petroleum are among the most useful of minerals. The Pacific Coast deposits are large, and are simply waiting the key of industry to unlock the granite doors and send them out into the commerce of the world. They, with kindred topics, shall be noted further along.

CHAPTER III.

VINEYARDS AND ORCHARDS.

IT is a law of nature that fruits reach their greatest perfection in a region having the most sunshine. Equable, or even monotonously mild temperature, like that of England, will not produce fruits in perfection, unless it is supplemented by clear skies and the fervent kisses of the sun, these, with the absence of severe cold in the winter, complete the conditions that make a fruit country; and it is a wonderful test and testimony of the wisdom which stocked the earth with capacities for man's use and benefit, that where these solar and other conditions appear in partnership, there the volcanoes have enriched the soil with the very marrow of the earth, and crumbling granite has added the strength needed to make it fitted for all fruit production.

In no part of the United States, and in no greater degree elsewhere over the globe, do we find so evenly poised these forces as in the New Empire, and hence it is that the whole coast, from the southern line of California to the slopes that guard Puget Sound, is of proved capacity for orchards or vineyards.

In Oregon and the North the apple is produced in the greatest perfection, and there the pear in all its luscious charms reaches a size and beauty seldom equaled.

In the East the only fruit to be relied on for crops is the

ORANGE GROVE IN THE NEW EMPIRE.

apple, and it has centers of the best production. The New York and Michigan apples will be the favorites in the market always, over the fruit of Ohio or Pennsylvania, though the apple grows from Maine to Minnesota, and the line of trees planted in the wilderness, from the Mohawk to the Ohio, by Apple Seed John, has expanded into thousands of orchards; yet the two States named hold to first excellence and their production is the favorite. This Pacific fruit region has the same peculiarity of premium locations, and Oregon has first place in the excellence of her apples. Though the Territories and Nevada and California have and will always crown many a feast with the wholesome fruit, the label, "Oregon Apples" will attract the consumer and bring the top of the market.

But it must be confessed that we might have apples and pears and plums in perfection elsewhere unknown and yet fail to attract any attention to the resources of our horticulture. Those fruits grow in every State in the Union, and while we might claim primacy in their production, we share it with all the rest. The unique value of our fruit region is, that it produces what grows nowhere else on the continent in any form or any degree in commercial volume.

In California is the center of the only genuine wine-producing area on the continent. Here and here alone are faithfully reproduced the conditions for viticulture which have made the wealth of Southern Europe. There five centuries have been devoted to the refinement of wine production. Here the origin of the industry and its rivalry of Europe are only twenty years apart. In that time we have developed hocks and clarets of defined character and as favorite with consumers as the Rhines of Germany and the Chateau Reds of France. In this wine-raising region there are choice spots, some of them already discovered and many yet to be. Here some thrill of sunshine, some subtle chemistry in the properties of the soil, or, may be, some balm in the breath of the winds imparts to the grape a subtle excellence. Who can tell what it is? When the quality of water used in tempering the steel of edged tools is proved to be the cause of excellence that no fire, nor forging, nor polish, nor skill can impart or take away, by what sensitive scale shall we measure the ineffable inspiration that gives varying excellence to the juice of the grape? This "spotted" tendency is one of the high evidences of excellence in a wine country. We drink the Johannisberger of commerce and call it good, and so it is, but the real Johannisberger is hardly known in trade, for its production is limited to a vineyard of only forty acres on the Metternich Estate in Nassau. Old Metternich, whose diplomacy set the European fashion after the downfall of Napoleon, knew nothing of wine-making, and one day ordered the stones gathered up and taken off the ground of the vineyard. That year his grapes made no wine, the juice sulked in the must. His superintendent told him that the wine had gone with the stones, and on restoring them the wine came back as excellent as before. Who shall tell why? All around, the same sort of stones are in other vineyards, but the wine is inferior. Who can tell why a few acres at Dijon should produce a still wine as mild as a dew-drop to the taste, yet with a subtlety of spirit that gives it almost the action of absinthe? And why do vines at Rheims, so few that they are numbered and counted, produce a cream wine, for which the gods of Olympus would have quit their nectar and the Wassailers of Valhalla their mighty mead?

Already it is noted that the same grape, planted in different localities, varies in its qualities, and, finally, California will offer as great variety in the individuality and excellence of its wines as Europe. Over three hundred varieties of grapes are grown here, and careful experiment is transplanting new sorts continually, for which the whole world, from Peru to Algiers, is put under tribute.

All this region of fruit tree and vine is dotted with orchard and vineyard. The statistics of horticulture are difficult of collection, because while you are reducing vine and tree to a census, enterprise is pushing the frontier of the area they occupy, and the counting is never done. In Egypt where the date palm furnishes fuel, food and timber is not of rapid growth, and is cherished by its owner, as the Bedouin guards his horse, every tree is tagged and taxed, and its age officially preserved.

The apple trees in California number about 2,500,000; in Oregon, 1,500,000; peach trees in California, 2,000,000; in Oregon, 50,000; California has 50,000 fig-trees, 2,000,000 orange, 250,000 lemon. By the census of 1880 the total fruit product of Oregon reached a value of $547,000; of California, $3,000,000. In the latter the value has, since then, doubled, and the capital invested in the fruit interest of California is $50,000,000.

For the apricot of California the demand constantly presses the supply. In its fresh state it is admired amongst the early fruits shipped East, and as it resembles the pear in ripening after it is picked, it is a valuable shipping fruit. Canned or dried, it has the world for a market, so it is not surprising that orchardists in this State have realized as high as $1,500 per acre for this crop. The nectarine is a luscious hybrid, which joins the best qualities of peach and plum, and is here more perfect in quality, and bears greater crops than elsewhere on the continent.

The dried fruits of this region have long been famous for their flavor and quality, and every year adds to the range of their market, and the refinement of the processes by which they are prepared.

The raisin industry of California was founded in 1872, when about a thousand boxes found their way to market. In 1881 the production had risen to 160,000 boxes, and in 1885 to about a half million. Our sunny and rainless climate, the tendency of our grapes to develop sugar, all join to make this industry one of our standard activities. Here an area of 20,000 square miles is specially adapted to producing raisins. It is a common experience of vineyardists to secure a net profit on their raisin crop of $250 per acre, and 20 acres of raisin grapes bring an income that cannot be wrung from a half section in some of the Atlantic States.

The wine yield of California varies with the seasons, but now seldom falls below 10,000,000 gallons, for which there is a ready and appreciative market.

OLIVE GROVE IN THE NEW EMPIRE.

Our brandy product is increasing, and some brands distilled here have beaten the world in a competition hotly contested. But why dwell longer upon the horticultural superiority of a region which tells its own story; where the currants are as large as cherries elsewhere grown, and the cherries are as large as the plums of the East; where all the stone fruits, and the pomegranate, the melon, and the Japanese persimmon, which grows to the size of an apple, and is called "The fruit of the gods," the custard apple of Burmah, will grow side by side,—a land where the orange, lemon and lime ripen their fruit clear up to the 41st degree of latitude; where the olive is gracing the hills and throwing its oil in commercial quantities from San Diego to Sonoma,—such a land wears its certificate of excellence as an outer garment to be seen of all men, and not as a hidden grace to inflame the fancy by a chance disclosure.

In all this area the English walnut and the almond, the leading nuts of commerce, grow as if indigenous, and our walnuts and our wines are kissed into perfection by the same kind sunshine.

Difficulties are besetting the centuries-old orchards, vineyards and olive groves of Europe. The world must more and more resort to the produce of our virgin lands for its supply of these articles, which fill so formidable a place in the diet and trade of the world. Our own country calls for such supplies for 60,000,000 of people, and did they all resort to us now the demand would doubly overgo our own capacity to supply. In these briefly narrated facts we see the solidity of the horticultural interests of the New Empire. Was there ever such a land since Moses looked upon virgin Palestine and found his fancy enchanted by its spreading meadows, its dewy vineyards, the yellow wheat that gilt its plains, the beauty of its flowers, and the plenty that its generous soil gave up to its people?

Here we have all this, and flocks and herds, with gold and silver mounted mountains standing guard over the wonders of the land; while the fish-filled waters babble their boast of rivalry in the task of feeding millions without calling for a miraculous draught of the nets.

HOMES OF THE CLIFF DWELLERS.

CHAPTER IV.

MONEY, ITS NATURE AND USES.

"Put money in thy purse."—SHAKESPEARE.
"The love of money is the root of all evil."—THE BIBLE.

NO people have a greater practical or sentimental interest in money than those of California. The discovery of precious metals in this State wrought a financial revolution in the world. It prevented the demonetizing of gold by our own Congress; and by greatly adding to the coinage of that metal, affected values to a greater stability, and wove the brilliant dream of Jackson and Benton into realization by making possible the payment of Government debts in " * * gold, yellow and molten, hammered and rolled." The silver store, long dependent on uncertain Mexico, got a like impulse from the discovery of the Nevada and Arizona deposits, and together the two metals have made California to drive out of fancy and even from fiction, the figure of Golconda and of Ormus and of Ind. Ophir, that gave up the gold that gilt the temple, is a tradition. California is an enduring fact, and from her mother lode and its ribs for generations to come, hardy miners will be digging and milling gold to make money for others to spend. By use of this store, generously poured into the world's lap, wars will be fought, soldiers paid, statesmen bribed. It will enable States to broaden their phylactery by extending their boundaries. It will arm and man navies to test the dominion of the sea. It will be the motive that sends men into the wilderness to redeem and plant the globe lands, that for the crops grown, they may get money. As the bride blushes at the altar, not the least pleasant anticipation of the delights of her new estate is the jingle of the coin her husband will throw into her lap. So, too, amongst the sorrows of the son as he buries his sire, there will creep in enough mental arithmetic to figure up the value in chinking coin of his share of the family estate. Every cradle, robe and shroud represents money, and dollars and dimes are woven into the suit of canvas spread by the ship to propel her through the waters, and they are in the anchor that shall hold her fast when the voyage is ended. Without money, commerce would be a rude barter, life would be without elegance, industry without variety, man without a motive. Where was the genesis of money, of a medium of exchange and a standard of value? David A. Wells has fancied it in his "Robinson Crusoe's Money;" but without refining upon definitions, money is anything into which you can convert your surplus labor and with which you can procure the surplus labor of another. A season's toil has produced wheat enough to feed you and some to spare. Another man has produced enough wool to make his own clothes and some to spare. The overplus in each case is the surplus labor. So A sells his wheat, that is, converts it into something that will buy B's wool after it has been transmuted by manufacture, and that something which enables this exchange of surplus labor is called money. No matter when it originated, if the world were swept by a besom to-day, and repeopled by a primitive race to-morrow to whom we would be as dim, distant and mysterious as the mound-builders are to us, that race would follow our foot-steps in commerce and the evolution of finance, because it would have the same wants and appetites as we have, and instinctively seek their gratification by the same means that we have used. In the cuneiform writings left by the stylus of many a Babylonian scribbler, we find the history of the Babylonian Rothschilds "Egibi & Son" who discounted notes, drove bargains and loaned money to the king, long before Abraham lost faith in the wooden gods he had whittled out with his jack-knife, and while Greece was a blank, Rome was a resort of coney-hunting savages, and what is now Europe was less known than we know the planets which are our goodly company whirling around the sun in inferior and superior circles.

As one after another the wheat-raising regions of the world have been subdued to tillage and have poured their crops into commerce, the result has furnished ground for speculation by political and social economists and financiers. So the corn and cotton belts of the world, as they contract by exhaustion or expand by discovery and experiment, are the objects of enduring attention.

But the products of these regions and the industries they support, differ widely from the precious metals, because they fluctuate and their value ebbs and flows, while gold and silver are the measure of that value. The moment they are freed from impingement with the baser substances in which they grow, they are *value*. They don't shrink and swell with plenty of famine in any part of the world. They are always in good demand, equally prized, equally desirable and equally capable of benefiting mankind.

After his Italian campaign, the great Napoleon was accused of sacrilege because in looting churches, his soldiers had despoiled shrine and altar of the images of the saints cast in silver and gold. Taxed with this, the Corsican answered that he had melted the saints into money in order that they might go up and down the world doing good as was the duty of saints.

As the precious metals have their inalienable qualities and functions, how much more should their production attract the attention of economists and financiers than do the perishing crops whose value they measure? So fixed has the public heart become in favor of gold and silver money, that we tear the sounds and sense and orthography of our language for terms to distinguish paper currency that is not redeemable in coin; "Shin-plaster" and "Red dog" are some of the names by which such currency has been known. Paper money, to be of genuine utility, must be of representative value and convertible into coin of one or the other of these metals produced by our mines. The precious metal product of this coast has given to our region its pre-eminent position. When an ex-premier of England stood before the Dons and Proctors of Edinburgh University for inauguration as Lord Rector, he opened his address with a figure in which California was used as a synonym for wealth. The non-exhaustion of our mines, the discovery of new processes which, by hitching chemistry and mechanism together, attack and reduce refractory ores or so cheapen the reduction of low grades as to make their working profitable, has had the effect of steadily maintaining our bullion output and convincing the world of the practical inexhaustibility of our deposits.

What eye has penetrated the depths yet under the feet of the deep miner? The earth is yet to be bored to greater depths before we go as far as men have gone for coal, and when we have sunk the shafts there is every assurance that the result will prove the region to be like a good watch, full jeweled in every hole.

In gold and silver we still lead the world, as official estimates in another column will show.

Russia regards her gold fields as the apple of her eye. The resources of imperial science and of imperial tyranny combine to urge them to the highest production, and yet, with an almost languid attention to our mines, we lead both her and England. In the thirty-five years preceding 1874, which includes the greatest output of the California and Australian mines, the world's yield of gold averaged $96,000,000 per year; so that with the comparative inattention to our mines here and the measurable withdrawal of interest in those of Australia, the yield fell only $14,000,000 short of the average of that period in which the placers were yielding their nuggets and dust to the rocker.

CHAPTER V.

GOLD AND SILVER.

THE romance of gold and silver mining is one of the most alluring chapters of the world's history. Gold and silver have stood in all literature as the synonyms for desirability. The Spaniards encouraged Columbus, the Genoese sailor, to embark in the experiment of seeking a new route to the Indies, because their fancy was inflamed by the vision of great spoil in the precious metals. With him, the incentive was scientific; he was hungry for geography. They wanted gold. When his voyage to India, sailing west to reach the east, was interrupted by an unknown continent, his followers and the Government, under whose patronage he was protected, began the hunt for gold in the new country.

To find gold was the hope of De Soto, of Balboa, of Cortez and Pizarro, and gold and silver soon loaded the Spanish galleons and they, in turn, were hunted on the high seas by British war ships in what amounted to actual piracy, though dignified by the name of war.

The desire for gold was soon planted amongst the foremost motives of our English-speaking people. It enlisted the pens and tongues of statesmen and economists. It affected profoundly the financial policy of this Republic, and to-day the issues that arise in gold and silver, their relative volume, the extent of their coinage, their intrinsic ratio, swallow up all other public questions.

The discovery of gold in California populated the coast more rapidly than would have been possible by any other means. If men had been promised immortality as a reward for the pains and perils of the journey here, they would have risked a refusal. But the temporalities promised by gold were an irresistible temptation. All of our other means of prosperity, our fields, orchards and vineyards, our wine and oil would now be the unproved elements in the clods of our valleys, had not gold brought to us a population in whose needs and industrial evolution were the germs of these great cognate productions. The consumption of the precious metals has kept pace with their discovery and production. There are placers yet unworked; there are quartz veins yet undeveloped; there are billions of gold yet to be mined from the Rio Grande to the Yukon.

The gold and silver we have already produced has reclaimed the Pacific side of this continent. It has been the cause of all other foundations, of agriculture, horticulture and manufactories. It has built cities, dug canals for irrigation, dredged rivers for navigation and constructed the lines of transcontinental railroads. It has established steamship lines, created commerce, wrested islands from barbarism and redeemed hundreds of thousands from poverty, and established them in comfort within easy reach of affluence. Nearly four billions of these metals have been taken from the mines, but the work is hardly begun. Using the experience already gained and applying it to ground untouched, to ledges unworked, there are billions yet to sparkle in the sunlight and jingle in the pockets of the people, to fill national treasuries, to turn the wheels of manufactories and spread the sails of commerce. The next thirty years will more than triplicate on this coast and in this country, the results of mineral wealth taken from our mines. This wealth is not to be shrunk by tariffs; it is not corroded by rust; it defies the gnawing tooth of time; even when the robber steals it there is only a diversion in its direction, for, through him, it again reaches circulation and fulfills its mission. For this coast the production of the precious metals means everything. It invests an idle population in an enchanting and profitable pursuit; it diverts labor from the production to the consumption of food, and makes better prices for the yield of the husbandman.

CHAPTER VI.

CURBSTONE BROKERS.

NO good thing can exist without drawbacks. The sun and warmth which generate the luscious flavors of fruits, give life also to the insect which preys upon the tree. Health and strength tempt to those excesses which bring both to untimely shipwreck. Even love walks lightly through the bowers in whose shadows jealousy, the counter-passion, gnaws its heart. The rich mineral resources of this coast which it would seem should always have brought wealth to those hardy and adventurous men who search them out, have proved the ruin of thousands through those parasites called "Curbstone Brokers." The prospector, developer or owner of a mining property appears in San Francisco and falls into the hands of a broker to whom the worth or worthlessness of the property is a secondary consideration. His method is as fixed as pocket-picking. By manipulation and coaxing, playing upon avarice and cupidity, he schemes for the control of the "mine." Getting a bond to cover it, he organizes a company and proceeds to capitalize the property, issue stock, and squeeze margins out of it by means foul or fair. One of these leeches went to a lithographer for a book of blank certificates of stock. "Where is your property located?" asked the artist. "'Damfino'—it's no difference anyway," replied the teredo of the street. "What is its name?" "Oh call it anything you like, but hurry up; I want to get the stock off." On the banks of Newfoundland to which the cod-fishers resort, no bait is wasted on the hook; a few scraps of white-fleshed clams are cast overboard and the fish rise to the feast, when the hooks, ragged out with strips of white cotton cloth, are cast in and taken by the fish, which are soon flopping on deck. So do these fishers of men who wring sorrow out of mines that should yield only satisfaction, profit, and an access of life's pleasures. Into the waters of speculation were cast the profitable Comstocks and consolidated properties which, by temporary investment, raised thousands to affluence; but they were followed by the rag stock certificates that represented a partnership of cupidity and scoundrelism, but were snapped up by the crowd that made no discrimination between reality and unreality. Your broker of the curb usually represents himself as with money, position and influence, or he has a circle of old-timers who are his rich friends, whom he inspires with confidence. He knows Jones, Flood, Mackay and Fair, who look to him for avenues in which to invest their money. Under these and kindred representations, a bond for a term of months is obtained, and then the time passes, weeks melt away. They are seeking the right men for directors; their wealthy clients are very particular; and, so the smooth lie runs, while the owner of the property, buoyed for a while on expectation and fed on falsehood, finds his expenses eating into his pockets and delay eating into his heart. Finally lying fails to explain the delay. The victim breaks the meshes of his net and investigates his way to the discovery that the broker is a penniless adventurer, as poor in morals as in pocket, whose only capital is falsehood, and his only merit the mastery of the art of persuasion. The owner came to the city to sell a mine. He discovers that he is sold instead. Now begins a struggle to recover what has been practically stolen from him. The broker insists that he has put money into searching for a market for the property. He has claims, liens and offsets against his undischarged trust under the bond. Put to legal proof he at last threatens to prevent the profitable capitalization of the property by any one else, using to that end his "influence on the street." If the owner quits him finally, the pest does not quit the owner, but spies upon his movements dogs his daily walk, and by every art known to criminal financiering, works to prevent the legitimate placing of the property where it would return solid dividends upon the hope invested in it. So the mining interest is hampered, the bullion output is limited, and the reputation of the stock and the fair name of the city's financial standing is tarnished by these illicit and immoral brokers. Many of them are broken-down politicians who began at an honorable elevation and fell from one treachery and broken trust to another, till they were dumped on the streets to live by their wits. They are the companions in declension to the women who begin in the roses and raptures of vice, the illicit pets of lascivious luxury, but who fall step by step to depend upon the occasional spoliation of the stranger picked up on the street.

True, amongst the brokers are men of untarnished honor, forced by circumstances to a repulsive association; but this inconsiderable leaven cannot make the whole lump wholesome, nor minister a cure of the damage inflicted upon our mineral interests by the vicious members of their guild.

There are unworthy men in all callings, and even the learned professions are not free from them. Law and physic shelter pretenders, and even in the house of God, the well-disguised hypocrite may break the bread of life for a time undetected. The criminally inclined seek the company of the virtuous as ambush for their designs, and so it has passed into a habit when a banker defaults or a fiduciary agent disappears with trust funds, we ask involuntarily, "What Sun-day-school did he superintend?" Emancipated from the bad name given it by operations of the ballooning broker, quartz and placer mining has a great part yet to play in the permanent prosperity of this coast. By refinement of processes, thousands of acres of placer ground will give up their treasures, and refractory ores that have defied reduction will yield at last, and many a block of stocks now hidden in forgotten places and thought to be worthless will enrich the owner who has damned and forgotten his investment.

It is not only personal thrift, but is the sign of good citizenship, to foster the legitimate rewards of our mines; while it is only good morals to chase the sinister and lying broker off the street.

CHAPTER VII.

DISCOVERY OF GOLD.

FROM the date of the discovery of gold in California, mining has been steadily and for many years rapidly gaining public confidence, and now it is justly regarded as a legitimate, safe business—one of the most important industries of civilization. When that discovery was made, the great mass of the people were incredulous, regarded the announcement as a sort of "Arabian Nights" or the tale of "Aladdin's Lamp" revamped and published under a new title. It is a historical fact, however, that since that day California has yielded $1,000,000,000 in gold. Astounding as this statement may be, it is strictly true. The same incredulity was manifested about the discovery of gold in Australia, but, notwithstanding that, the island continent has produced £200,000,000 in gold, and the grand total productions of bullion from all sources since 1848 amount to $5,862,165,000.

When the silver mines of Nevada were discovered, no one believed, not even Comstock, the discoverer, nor even contemplated the vast treasures stored away in those inhospitable mountains; but Nevada has sent out $350,000,000 in silver, and the production of the United States since 1858 has

SITE OF THE DISCOVERY OF GOLD.

been $776,780,670. Then came Colorado, Montana, Idaho, Utah, Arizona, and Dakota, each having to stem the tide of popular distrust, to prove by actual demonstration the existence of precious metals in their river beds and mountain ranges; but they have given their millions to the commerce of the world. The total bullion yield of the new empire since 1848 reaches the enormous sum of $2,607,006,786.

The early pioneers, the brave, the intelligent, the industrious pioneers of those mineral regions, were invariably looked upon by the incredulous as voluntary exiles, sacrificing home, friends and the comforts of civilization for the wild life of a frontiersman. But the results, the magnificent results which these pioneers have achieved, the victories they have won, and the long list of those who now count their coin by the million, gathered from these newly-found mines, is a proof which the world is compelled to accept of the wisdom of their course.

In the face of this enormous yield of precious metals and all that has been achieved through this yield, much is said by a certain class about the money that has been expended and the losses sustained in mining enterprises. True, mines have been purchased at almost fabulous prices, but, in nearly every instance, when the purchasers exercised the same judgment that careful business men would use in other transactions of equal magnitude, they have received rich returns for their investments.

During the year 1884 there were 5,582 failures by those engaged in other callings in the United States alone, with total liabilities amounting to $81,155,932. This sum, the liabilities for one year in the United States alone, is greater than that of all the failures in mining enterprises from the landing of Noah's Ark to the present day, whilst the losses, through the failure of banking and other business houses in Europe and America, have been simply appalling. But this appears to be regarded by the anti-mining class as the legitimate effects of natural causes. Millions may be lost through corrupt bank officials, scheming railroad magnates, or those engaged in commercial pursuits, without apparently shaking their confidence or provoking a feeling of distrust; but, if a few thousand dollars are absorbed in a mining enterprise without returning at least double the amount invested during the first three months, the investor proceeds to get up a general howl, and whines and sniffles about it as though an irreparable calamity had befallen him.

But losses are not only sustained through the failure of banking and commercial houses and bankrupt railroads, but bankrupt States and municipalities. From a statement in a recent number of the *Money Market Review* it was shown that English financiers had advanced by loans to the several bankrupt States of Europe and South America upwards of £600,000,000 or $3,000,000,000 in twenty-five years, and at that time the market quotations of the stock gave it a value of a little over £60,000,000, so that in a quarter of a century there had been a depreciation or loss of over £500,000,000 or $2,500,000,000. These loans had been to Turkey, Spain, Greece, Egypt, Mexico, Grenada, Venezuela, Iquique, Honduras, Peru, Chili, Paraguay, Uruguay, and other places. Lord Derby, in a public speech some time back, stated that the loss to British capital advanced to defaulting States alone had been over £300,000,000 or $1,500,000,000. Although a considerable amount of the money loaned to those countries might have been re-invested in English goods, there can be little doubt that by far the greater portion of the bullion sent to these countries has become absorbed amongst the population, and the Governments, in most cases, are unable to pay the interest or principal. Nearly one-half of the new working capital of gold now furnished to and distributed throughout the world by the gold-mining population has been unfortunately sunk in these bankrupt States of Europe and South America. The same authority goes on to say, "Had the financiers and capitalists of England devoted a tithe of that vast sum so irretrievably lost in foreign bankrupt States, to the practical developments of gold-mining resources of the Australian Colonies, they would not only have materially aided the legitimate developments of mining, increased the supply of gold or purchasing power, and fostered other industries and forms of wealth incidental thereto, but would, in all probability, have been amply rewarded for their outlay."

Not only has mining produced this $5,862,165,000 of bullion, but it has wrested the Pacific half of America and the island continent of Australia from a wilderness. It has created a demand for new industries, has created these new industries and hundreds of millions of wealth in permanent improvements, furnishing employment for the labor of the

CONDUCTING WATER TO THE MINES.

poor, and the capital of the rich. Thomas Cornish, in an article published in the London *Mining Journal*, says: "The value of our gold supply has occasionally received attention at the hands of some writers on finance and political economy, but it is somewhat surprising that a subject of such vast importance to the general progress of the world has not been more fully dealt with." And continues: "There can be little doubt but that the unparalleled production of new wealth by the gold and silver mines, has been the primary cause of the rapid progress of events, the enormous increased wealth and prosperity of many civilized nations, and in consequence of this general advancement of wealth, intelligence, trade, commerce and finance, it has become an absolute necessity that the annual production of gold should not maintain its present standard, but that the supply of new gold should increase annually in the same ratio as trade, commerce, and population." Mining must, therefore, be considered one of the most important industries of the world, and one to which there should be more intelligent consideration given than has heretofore been done.

And, adding another testimony, "It will not be questioned," says Mr. Stephen Williamson, a conservative English writer, "that the large increase of the world's money, due to the Australian and Californian mineral discovery, led to a great extention of the world's commerce. The interchange of commodities was marvelously stimulated. Labor had for many years a greatly augmented recompense; the material comfort and welfare of mankind were greatly promoted. Real and personal property increased enormously in value all over the civilized world. The foreign commerce of England alone rose from £250,000,000 in 1852 to £650,000,000 in 1875, and it has been gradually increasing to the present time. The foreign commerce of many other nations rose in like proportion."

Production of the precious metals throughout the world in 1884:—

AMERICA.

	Gold.	Silver.	Total.
British Columbia	$ 3,000,000	....	$ 3,000,000
United States	40,000,000	$47,000,000	87,000,000
Mexico	1,000,000	15,000,000	16,000,000
Guatemala	2,000,000	405,000	2,400,000
Honduras	750,000	150,000	900,000
San Salvador	1,125,000	225,000	1,350,000
Nicaragua	875,000	175,000	1,050,000
Costa Rica	250,000	50,000	300,000
Columbia	3,000,000	1,000,000	4,000,000
Peru	1,000,000	5,000,000	6,000,000
Chili	1,000,000	3,000,000	4,000,000
Buenos Ayres	1,000,000	1,000,000	2,000,000
Argentine Republic	1,000,000	1,000,000	2,000,000
Brazil	2,000,000	1,000,000	3,000,000
Other Countries	1,000,000	1,000,000	2,000,000
Total	$59,000,000	$76,000,000	$135,000,000

EUROPE.

	Gold.	Silver.	Total.
Russia	$13,000,000	$1,000,000	$14,000,000
Austria	2,000,000	1,000,000	3,000,000
Prussia	1,000,000	1,000,000	2,000,000
France	1,500,000	2,000,000	3,500,000
Spain	1,000,000	1,000,000	2,000,000
Other Countries	1,000,000	1,000,000	2,000,000
Total	$19,500,000	$7,000,000	$26,500,000

ASIA.

	Gold.	Silver.	Total.
Japan	$1,500,000	$2,000,000	$3,500,000
Borneo	3,000,000		3,000,000
China	2,000,000		2,000,000
Archipelago	3,000,000	5,000,000	8,000,000
Total	$9,500,000	$7,000,000	$16,500,000
Australia	$18,000,000	$1,000,000	$19,000,000
New Zealand	7,000,000	1,000,000	8,000,000
Africa	4,000,000	1,000,000	5,000,000
Oceanica	1,000,000	1,000,000	2,000,000
Grand Total	$118,000,000	$94,000,000	$212,000,000

Now we have found by this investigation that the production of bullion since 1848 amounts to $5,862,165,000.

That mining during the last thirty years has created more wealth, stimulated greater enterprise and industry, and produced more beneficial results to the commercial world than all the other industries combined.

That mining is a safe, legitimate business when conducted on sound business princip'es.

That the hazard and loss to capitalists are less than in most other enterprises in which men engage and invest their money.

That the profits derived from mining are larger and more regular than in most avocations in life.

That the demands for the precious metals are increasing year by year, and that their continued production is a paramount necessity.

That mining is deserving of and should receive the attention of scientists, financiers, and the enterprising men of the world

CHAPTER VIII.

DEALING IN STOCKS.

There are three ways in which property can be rightfully acquired:
By labor, which includes legitimate speculative investment.
By discovery,
And by voluntary gift, which includes inheritance.

THERE are but few, if indeed there is a single question of any magnitude of a public nature on which all agree, and it is right; it is natural that thinkers as well as the unthinking should differ, because from such difference much good issues. Mining, for instance, in all its phases, is represented by some as purely a business of chance or speculation; thus conveying, or attempting to convey, the impression that this branch of industry is more hazardous than most other undertakings where industry and capital are necessary to abundant success. But the doctrine is absurdly erroneous. In point of fact, when reduced to its proper standard it will be found that all monetary success may be summed up in that one word, speculation. Look at it, turn it, analyze it as you will, the speculative element is blended with all our secular affairs, pervades every business avenue of life. Vast fortunes have been amassed in every quarter of the civilized globe, but by whom—from what particular business? Not necessarily the high born nor those of scholarly attainments or accomplishments, but to the speculator in speculative ventures, to those who grasp the present, forecast the future, and discount results; the men who resolutely embark in large mining ventures, who invest judiciously in real estate, who connect themselves with and manage great railroads and railroad enterprises, or who engage as wholesale merchants in goods of universal necessity,—these are the men who amass colossal fortunes. To accumulate wealth as the miser does, is a slow process indeed, and will not compare in its results with the grand operations of bold, yet prudent men. The man who determines to invest in real estate, selects his location in or near some prosperous and growing city; for years his investment may not seem to pay, may indeed be a burden to him, but when, by the natural growth of the city and surrounding country, his lands are enhanced, he finds himself rich as by magic. The projectors and managers of railroads move with greater celerity, if with less certainty. Their success depends less upon the efforts of outside parties, and more upon the vigor and persistence with which they push their own enterprise. Again, the control of large sums enables them to exercise their financial abilities in many channels at the same time. Their connection, moreover, with a certain clique, makes them possessed of everything worth knowing regarding the market position of stocks, and here is really where great fortunes are made, and made quickly.

Vanderbilt, Gould, Sage, and their compeers, spent years of early life in accumulating what in late years they would realize in a single day. Stanford, Crocker, and their associates, are exponents of the same doctrine. The merchant's gains come more slowly. Great competitions circumscribe the profits of all small dealers. It is only the wholesale dealer, the man who ventures largely in staple articles, that can hope to rise speedily in the scale of fortune. The pork packers of Cincinnati and the grain merchants of Chicago present examples of the speedy acquisition of wealth by large and quick transactions. But John P. Jones, James C. Flood, John W. Mackay, James G. Fair, and George Hearst afford still more brilliant examples by the princely fortunes extending into the millions which they have amassed in the mines of the Pacific Coast. A few years ago they were poor men. They examined the situation; they forecast the future and operated boldly for grand results. They stepped out of what is known to the slow plodder as the rut of legitimate business, and entered upon the domain of speculation, honorable speculation, and all true speculation is honorable.

The same opportunities, the same line of action pursued by them, is open to others who have the dash and the moral courage to emulate their example.

But it may be urged that all do not make money who engage in mining; neither do all succeed who engage in banking, in mercantile pursuits or any other profession or calling to which men direct their ability and energy. Prosperity and adversity are to be met in all the diversified walks of life. The prompt business man with judgment and nerve may fairly expect as large and rapid returns from the investment of capital and the employment of his time in sound mining transactions, as in the most promising, in fact as in any other avenue now open. But it may be said that mining leads to stock-gambling; so do canals; so do railroads; so do Government securities; so do all enterprises too large for private undertaking and doing business necessarily in the form of stock companies. But shall we therefore have no more canals, no more railroads, no more great undertakings of any kind, because, as it is urged, unprincipled men seize upon mining stocks as a favorite means of gambling? Shall we, therefore, abandon sound mining and mining transactions? As well abandon money itself, because in what profession, trade, or calling are there not gamblers, schemers and workers of iniquity? Do not journalists prostitute their journals for gain? Do not lawyers sell their abilities to any man who has the money and chooses to pay them? And do not clergymen have the loudest call to parishes which pay the largest salaries? Do not some merchants give short weight and adulterate their wares? Do not some milkmen extend their fluid at the town pump, and whisky men enlarge their supplies from the kerosene can? Do not some cloth and paper merchants know the value of shoddy and sizing, and so on to the end of the list? It is idle and mischievous to select the members of any particular class or profession, of generally respectable people, and attempt to make them out worse than their neighbors. Human nature is human nature all the world round, in all grades of society. "Cast out the beam out of thine own eye, and then shalt thou see clearly to cast out the mote out of thy brother's eye" is a short and pungent exhortation that all will do well to remember.

The annual yield of our mines is a proof positive of their excellence. The enormous fortunes made during the past twenty-five years by successful dealers in mining stocks, are witnesses, the power and force of whose testimony neither sophistry can weaken nor argument overthrow. They are realities to be measured and counted by all. Reduced to a business basis, whether in stocks, gravel beds, or quartz ledges, mining is precisely like any other business, with its bright and shady sides, in the main just what those engaged in it make it, nothing more, nothing less. "From quack lawyers, and quack doctors, and quack preachers, good Lord deliver us," was the prayer of the pious old farmer, and he might, with propriety, have embraced a few other quacks in his petition, in order, as Mrs. Whittlesey would say, "to make the platform broad enough to *kiver* the *hull* ground."

When it becomes necessary to transact business, whether in stocks or otherwise, through the medium of an agent, it is wisdom, it is safe only to select a responsible, reliable man, one who is conscious of the fact that his business success depends on his doing right.

CHAPTER IX.

OTHER MINERALS.

ACCORDING to the report of the United States Geological Survey for 1884, the enormous sum of $800,000,000 is invested in American mining enterprises, all branches included, as productive capital; nearly half a million people are employed, and the annual production for the period over which the report runs was $413,104,620, or over fifty per cent. on the capital invested. Just drive a peg there.

At the first blush these figures may be regarded with astonishment by a large number of generally well-informed people. As the figures are official, however, they must be accepted as correct.

But we are particularly examining the mining interests of the Pacific slope of the continent, all acquired territory, and most of it long after the nation's independence. California, the key-stone of the industry, practically came into the Union

CAMPING IN THE NEW EMPIRE.

on the 7th of July, 1846, when the emblem of liberty and
progress was hoisted at Monterey, and gold and silver, the
most valuable of all metals, are chiefly being considered.
The gold is found in two general divisions—placer and
quartz—and a large number of men are engaged in both
divisions. The pan, rocker, flume, sluice and other methods,
including hydraulic power, are common on placer fields; but
in quartz mining expensive machinery is necessary, and also
a higher grade material. Although a few placer fields have
been discovered in different sections of the coast, those of
California are by far the largest, richest and most enduring;
they are thousands of acres in extent, and their product is
counted by the billion.

SILVER.

Nevada may fairly be styled the *alma mater* of silver-
mining in America, and, indeed, the world, for she excels all
other regions of equal radius, in production, and has been the
educator of the world in the silver-mining business. Up to
the discovery of the Comstock lode, the methods in vogue
for taking out ore and extracting the silver were crude in the
extreme. All the valuable new processes discovered and
applied, through science, ingenuity and skill in this art, date
from Nevada; and she has graduated a long list of brilliant
men, given them fortune and fame, and set them as lights on
the mountain-top, to guide others of equal courage, industry
and frugality, to equal fortune and equal fame.

A great deal of fault has been found with the management of these mines. Well, the men who had control of
them doubtless managed them to suit themselves, just as
those engaged in other avocations managed their business,
and the privilege is open to these fault-finders to manage
mines for themselves; they can either buy or go out into the
mountains and discover. But they have neither the money
to do the first nor the courage nor industry for the latter;
their cry is, "*Divide.*" The world is full of tramps, socialists,
renegades from good families, honorable professions and excellent opportunities, who find fault with everybody but themselves; whereas, they alone are to blame for unappreciated,

wasted talent and unappropriated opportunities. But this is no reflection on the mines, or the business of mining, nor does it detract from the value of the $350,000,000 given up by Nevada's mineral lodes, nor the millions annually produced by the mines of other Pacific States and Territories.

QUICKSILVER.

Quicksilver, next in value and importance, is also found in the Pacific empire; but, to more particularly localize the deposits, they are found in California, and within a hundred miles of San Francisco. In gold and silver mining it is an indispensable requisite; without it neither could be carried forward successfully. There are in all about fifty so-called quicksilver mines in the State, but only a dozen that have developed sufficient merit to deserve the name, with a total production of 5,500 tons. The yield for 1885 was 32,073 flasks of 76½ pounds each. During the past twenty-five years the price has ranged from 35 cents to $1.55 a pound.

COAL.

The production and consumption of coal on the Pacific Coast are increasing. In 1883, the amount brought to San Francisco alone was 899,301 tons; in 1885, it was 1,223,339 tons. Of coke the importation in 1884 was 10,695 tons; in 1885 it was 20,611 tons.

Coal mines are found in California, Oregon, Utah, New Mexico, Arizona, Colorado, Sonora, Washington and Alaska, and in British Columbia, the last-named place furnishing the best quality. But the importations are from Europe and Australia as well as from British Columbia, brought here by vessels in ballast, a better article, with cheaper transportation, than from our own collieries.

Our coal deposits are numerous and extensive, but so far as discovered are of the bituminous quality. They are inconveniently situated, and are burdened with heavy transportation. Three wealthy corporations are struggling for the control of the market; they have the money, business experience and appliance, hence it is unlikely that any new mines will be opened for a long time to come, as it would be entering a hazardous field already monopolized, against dangerous competitors.

PETROLEUM.

Pennsylvania was long regarded as the coal and iron State of America; then she led off in the production of coal oil, and still continues to rank the world in this commodity. As in quicksilver, so in petroleum, California is the only State on the Pacific slope where oil is known to exist in paying quantities, or where any considerable money has been invested in the business. In Montana, Idaho and Washington, what are called surface indications have been found, but no discovery of consequence has been made, while the general formation and the broken nature of the country are thought to be adverse to profitable oil wells in those regions. As far back as 1865 capitalists became interested in the oil regions of California, and since that time they have spent a million dollars or more in machinery and in boring wells. The oil production of the State for 1885 was about 5,000,000 gallons, being an increase of 25 per cent. over the preceding year; and it has been demonstrated that this output can be enormously increased. But the oil, like the coal business of the coast, is practically controlled by powerful monopoly.

Iron, copper, antimony, lead, asphaltum, sulphur, soapstone, graphite, gypsum and diamonds are also found on the coast, and some of them in large quantities and of excellent quality; but capital, courage and industry are requisite to develop and gather their great wealth.

Beginning with the first gold excitement caused by the placer discoveries in California in 1848, between that year and 1865 we had: The California quartz discoveries in 1851; the Australian gold find in 1852; the Oregon gold excitement in the same year; that of Washington Territory in 1854; the Peru gold rush in the same year; the great copper discoveries on Lake Superior in 1855; the Arizona discovery in 1856; same in Nevada in 1857; the Frazer River rush in 1858; California copper in 1860; Pennsylvania petroleum in 1863; the Reese River boom in 1864; California petroleum and Colorado gold excitement in 1865—and the end is not yet. System and organization may prevail more largely than before, but let it be remembered that only a tithe of the precious metals has yet seen the light on the Pacific Coast. Many a bonanza is waiting for the lucky man, and many a ledge abandoned for a rush elsewhere, is biding its time for exposure of fabulous wealth.

CHAPTER X.
YOSEMITE.

THE Pacific Coast is rich in natural wonders. Even as you approach it from the East, the way is thick set with deserts over which the *mirage* shimmers with its disembodied forests, flowers and fountains, and the mountain gateways are pillared with grand forms of many colored rocks. The exalted fancy is fed upon these scenes which lie in front of the curtain which is finally lifted as the traveler passes the summit of the Sierras and slips down their hither slopes. He may have left winter behind in the Mississippi Valley, but here he is in the midst of spring. The forest around is vocal with the song of birds; the vineyards and orchards are offering their promise of fruit and wine; and with his glance resting on green turf and flower-spattered fields, high above them all, he sees the white line of snow resting on the Sierra's serried spine. The impression is never forgotten That chain of mountains is the Himalaya of California, and on the plains below flow the counterpart of the holy rivers of India, and only the lowly Sudra and the lofty Brahmin and the great gulf between them are lacking to make a Hindostan in miniature. Once within the mountain walls there are problems in botany, geology, zoology and mineralogy which excite the wonder of scientific men; but aside from these which invoke skill in chemistry, knowledge of vegetable physiology and comparative anatomy, we have in the wild scenery of our mountains, in our hot springs and geysers, a series of related phenomena furnished by no other part of the world. But after an industrious curiosity shall have seen them all, when, if possible, the senses are jaded by the unfolding marvels, then let Yosemite be seen.

GENERAL VIEW OF YOSEMITE VALLEY.

The route to it lies down the Southern Pacific railway to Berenda, where the just constructed Yosemite road leaves the main line and shortens the stage ride to the Valley by one day, and the only day on the original route that was listless with lack of interest. By the Berenda road the first day's dinner is at Grant's White Sulphur Springs, where Judge Grant, formerly chief justice of Iowa, founder of the smelting industry of Colorado, president of the National Trotting Association, and millionaire, has founded the most elegant of resorts upon waters that by contrast leave the famous White Sulphur of Virginia without virtue. From Grant's, on the splendid road, spurned by the flying heels of the six horses which pull your coach at an unceasing gallop, the charms and marvels of the mountain region multiply every moment. Springs fed by the snows that are still far above your head, burst from the rocks by the way-side. You drink the crystal water. It is nectar. Around and above you tower the sugar pines, on whose sides the crystallized sugar stands in pine-apple-like masses.

The elevation, the air clarified of impurities, the lilies that embroider the carpet of turf and pine needles, the frequent lofty outlook across the great San Joaquin Valley, across the dwarfed Coast Range and to the Pacific Ocean, whose surf roars and rolls the sand a hundred and sixty miles away, all join to transfigure the beholder and make him seem to step out of his former self, as he has in fact gone out of, and above that world in which he felt like a worm, while here he feels like a god.

At Clark's Mountain House there is rest for the night, and for breakfast apt to be mountain trout just delivered from stream so cold that the scaly beauty just out of it, pains your hand if you hold it. Such waters Izaak Walton never whipped with a fly, and such fish crisped over a broiling fire of cedar coals, no king ever ate. Near this Mountain House are the big trees of Mariposa, nearly 500 in the group, prone and erect. They stand, the survivors of the earliest vegetation that came upon these mountains when nature had done retching, and her upheavals were finished. Here they have

THE TWIN GIANTS.

stood in solemnity and majesty all their own, and the years counted back to their germination melt into a perspective so remote that it easily embraces the earliest recorded history, and includes the rise and fall of empires that were dissolved by old age when they had reached a thousand years. But to the monarchy of these kings of the forest there has been no end. They are not a dynasty, for nature has confessed her incapacity to repeat the effort which brought them forth. Without ancestry and without posterity, requiring the framework of a continent for their throne, they are the type of that eternity with which their sapling growth began and with which they endure, matchless, majestic, inscrutable!

Around their giant trunks and in their shadows, in his childhood, played the Indian Sequoia, and in his manhood he noted their difference in size, in foliage, bark and seed cone, from the neighbors which grew in their shadow; and as he was the first to guide to them German botanists, the trees that had sheltered his infancy, and had awed his dawning consciousness, were called by his name, "Sequoia."

Loth to leave the trees and the mountain house the traveler, whose roused interest has risen superior to fatigue, looks forward to the second and last day's stage, which is to end at the valley. He has a conception of it. A valley suggests to him browsing herds whose bells tinkle while they nip the herbage or chew the cud. Just how such a valley shall rest amongst these lofty mountains, grows into the speculation of the traveler as the coach maintains its skyward flight, and at last he seems abreast of the bald ridge, in all months snow-crowned. If his entry of the Yosemite be by Glacier Point, he finds himself looking out over mighty corrugations high above timber line, and with but little to encourage his cherished vision of a pastoral valley. Where can be the cascades and water-falls, the limpid river making a silvered line down flowery meadows? Around are rocks, massive, pitiless, ponderous, and immutable, whose crevices offer no home to grass, blade, or shrub, and which grudge the uncanny lichen its scanty living and home inhospitable.

Where is the valley? All at once, from Glacier Point, instead of looking ahead, you look down, and, straight as a bullet would drop from your hand, 3,500 feet beneath, lies the valley. You are standing 7,201 feet above the sea, and the verdant floor of Yosemite has dropped half that distance back. Here are no gentle slopes, for the granite walls that shut the valley in rise mostly as straight as the plumb-line can drop. No description can do justice to this greatest natural marvel in the world.

At Niagara you have the world of water droning an eternal doxology, then the gorge, the whirlpool, and, well, the hackmen. But here are the grand steps that lead up to the king of natural glories, and tell as much as pen may and tire. After it is seen, you realize that the half has not been told, for trees, mountain peaks, cañons, forests, and naked rocks piled in that confusion in which nature cast them as her work was finished, had made you expect to see something all unlike this focus of all natural wonders.

It is as if the topmost ridge of the Sierras, which here preserves a main direction of north and south, had been parted right across, been pulled apart when it was so plastic that the general shape of the wound it left is that of a birch canoe, wider in the middle and tapering at the ends. Upon the floor of this space, disintegrated granite has sifted from the walls, and the inflowing streams have brought the elements of soil until its general surface is a meadow, charmingly dotted with trees, and from the height at which you are looking, if the sun be streaming through the western cleft in the rocks, it looks like a *bijou* carving, like a cameo setting to a ring or brooch. But you are looking down upon a tract 14 miles in length by 3 broad at its widest part, and into it there pour water-falls that leap, unchecked, from heights varying from 500 to 3,270 feet. Here is Po-ho-no, the Bridal Veil, 860 feet, and looking like a white plume swayed by the wind; Yosemite, 2,548 feet, a great web of fluffy white satin, it seems, hung over the cliff; Pi-wy-ack, Vernal Falls, 336 feet, called the Cataract of Diamonds, because of the lights that play upon it; Yo-wi-ye, the Nevada Falls, full of splendors as the sunset strikes its snowy surface; Tu-lu-la-wi-ack, South Fork, 500

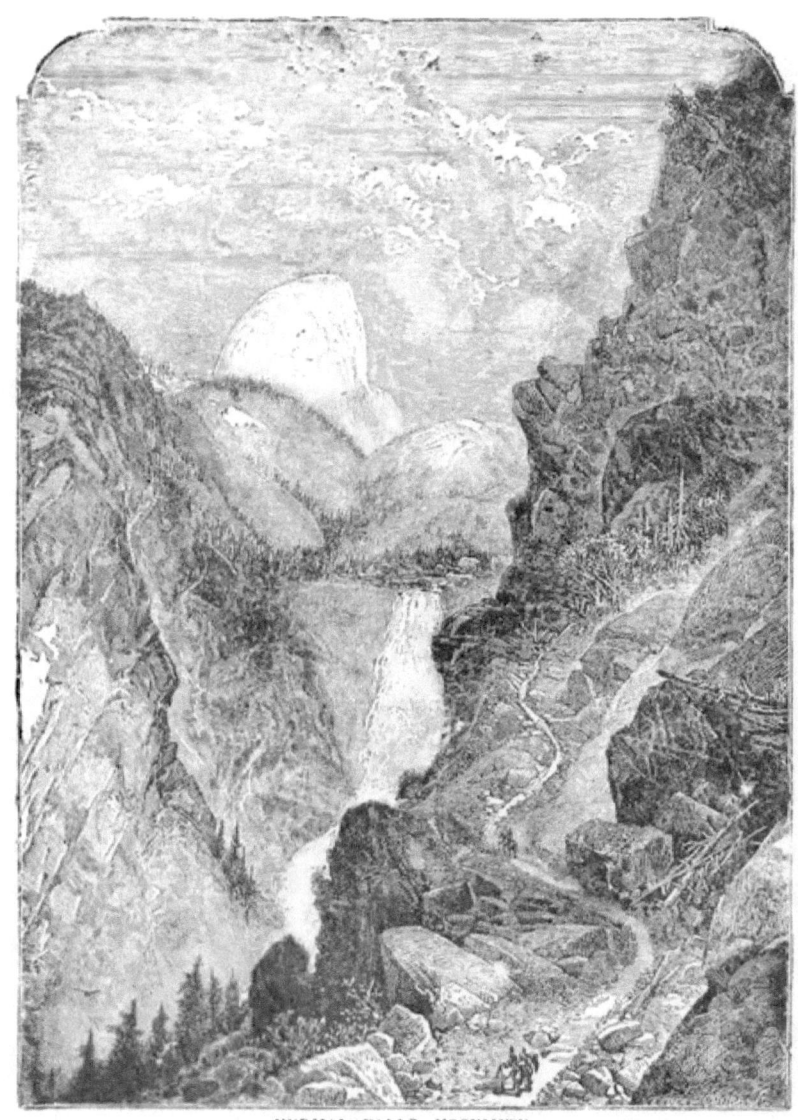

VERNAL FALLS, YOSEMITE.

feet, and Loy-a, the Sentinel Falls, 3,270 feet. Where else in the world has nature so used her fountains for grand effect? Standing at Glacier Point, you see the mountain peaks, whose feet are planted in the green turf of the valley far below. You are told their height, but you are dumb to all conception of its immensity. Here is Tis-sa-ack, the Half Dome, by the original Indian occupants fondly called the Goddess of the Valley, 8,823 feet high; Cloud's Rest, 9,912 feet; To-coy-ac, North Dome, 7,526 feet; Glacier Point, from which you took your first look, 7,201 feet; Cathedral Rock, 6,631 feet; Mah-tah, the Cap of Liberty, 7,062 feet; Mount Starr King, 9,080 feet; Union Point, 6,290 feet; Pom-pom-pa-sus, the Falling Rocks, or Three Brothers, 7,751 feet; Poo-see-nah Chuck-ka, the Cathedral Spires, 5,934 feet; Sentinel Dome, 8,122 feet; the Sentinel, 7,065 feet; Inspiration Point, 5,248 feet; and grand old Tu-tock-ah-nu-lah, chief of the valley, El Capitan, not so lofty as some, but with a certain broadness of shoulder and solidity justifying his headship of this mountain clan of mountains, rising 7,012 feet. Where again in the world has nature planted mountain peaks so thickly, and so distinguished for features that draw pilgrims from every country?

You are at Glacier Point; you descend by a trail that zigzags down the wall. You may do it on foot, or on horseback. When you are down in the valley, turn and look back along the way you have traveled; it is like looking against the straight white wall of your room. High above you appear other parties coming down the path you have just trod, and they look like specks stuck against a perfectly perpendicular wall.

On and into the walls on both sides of the valley these trails are cut, and men and women, mounted on sure-footed horses, go up and down like flies walking on the window pane. If your head and hand are steady, there is no end to the exhilarating adventure furnished by the cliffs and mountains that girt and guard the placid green of the valley. A Scotch bird-catcher, of St. Kilda, has scaled one of the loftiest peaks and planted an iron mast in its summit, to which a rope is fastened. By taking hold of this, with feet against the rock, and going hand over hand, you can pull yourself to a height of nearly 9,000 feet. Many try it, and several ladies have succeeded. But one can admire Niagara without shooting the falls in a canoe; and one can get an experience without which a life-time seems barren, by staying on the charming walks and drives of the valley's floor, or, at most, trying the safe trails.

When the moon is at the full and rises over the eastern end of the valley, so in line with the Nevada Fall that her silver seems to be pouring out of her face and down the cliff, sights are seen of such majesty and so full of inspiration that descriptive language is as idle as dumb show. As the moon climbs the sky, one side of the valley is in solemn shadow, while the white walls, peaks and water-falls on the other side take on a softness and tone that easily persuade the fancy that the eye is beholding something that is not of this world.

If you are not so lucky as to catch a full moon, it entails earlier rising that you may see the day born and pillowed in this cradle made on purpose for the young god. For quiet pleasures there are buggy rides, and then down through the valley's midst flows the crystalline Merced. Its waters are liquid diamonds, and floating over the pure white sand are rainbow trout that drive you wild. But after all this is said, and were it even said by the inspired tongue of an archangel, there comes the same despairing admission that no painting, photograph nor phrase that can be framed in words, can describe the beauty, purity, majesty and awe-inspiring grandeur of Yosemite. Hence all is focused in this advice: Take the train to Berenda, see for yourself and bring away an impression that will endure like the memory of your first kiss.

CHAPTER XI.

SCHOOL, PULPIT AND PRESS.

IT was a statesman who declared that, compelled to choose, he would rather have newspapers without Government, than a Government without newspapers.

COLLEGE IN NEW EMPIRE.

The press, pulpit and school are the unofficial, volunteer, spontaneous institutions of civilization. Its merits are measured by their excellence. They are the barometer, thermometer and wind gauge by which our moral meteorology is registered.

The schools of California were founded in the first constitution under which the State was admitted to the Union. The foundation was ample and it became the model for the other States and Territories of the New Empire. The opportunities for common school culture are as accessible and their scope as satisfactory here as in any part of the Union. The country school is up to the severe standard of New England excellence, and the city systems offer facilities for liberal culture unexcelled.

To illustrate, Oakland, the second city of California, has attached to her city school system an astronomical observatory, with telescope equal to that of Albany or Chicago, and a complete set of instruments for the study of physical and mathematical astronomy. No other common school system in the Union has such an adjunct. This splendid equipment entire, is the gift of a public-spirited citizen. The tendency of our people to encourage science and culture has many significant illustrations, as the Lick Observatory, the library and art gallery of the Berkeley University and the foundation of the Stanford University at Menlo Park.

The thoughtful parents planning a migration and the foundation of a new home, always ask first, "What are the school facilities?" Here under the gifts and endowments relating to higher education, is the ample foundation for the common school, the origin and source of that fundamental knowledge which is absolutely necessary to a contented and prosperous life, and which in all cases is the thread to be followed forward into ampler learning and upward to the highest attainable culture.

The tinkle of the school bell follows the 9 o'clock sun across the New Empire from the eastern line in the mountains to its western border that slopes into the sea, and all along rise the chaste walls of seminaries, colleges, convents, schools and other institutions existing by private enterprise or the patronage of different churches; and as the sun slopes to the west, the future rulers of the New Empire troop from the

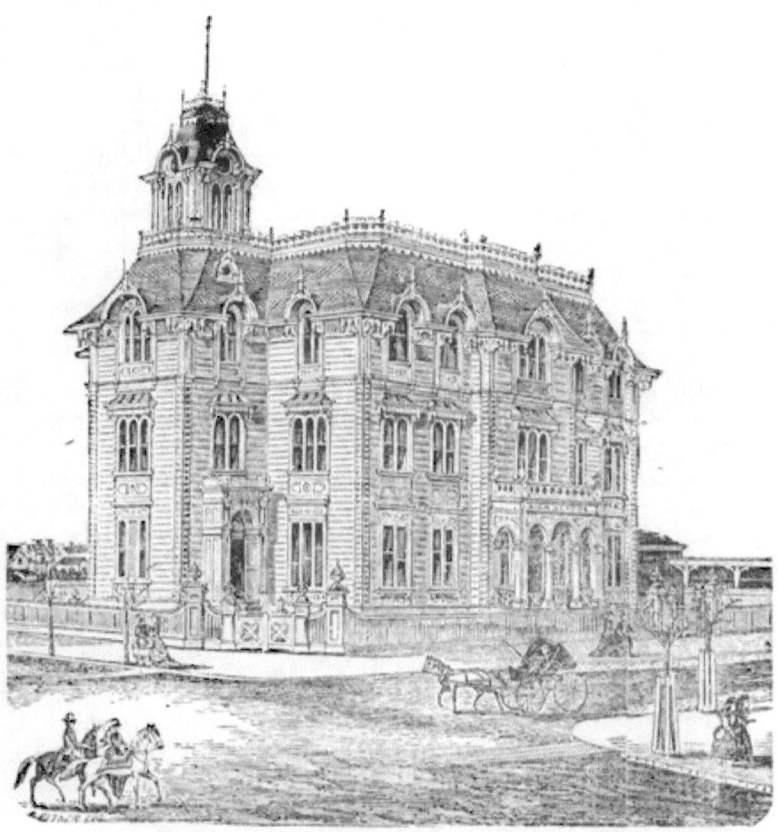

OAKLAND HIGH SCHOOL BUILDING.

school-house door homeward, and as 4 o'clock rings along, leaping the meridians of longitude like a race-horse taking the hurdles, this army marches, shod or shoeless, still in its discipline and brain and brawn girt with the hopes, the happiness and the greatness of all the future. Beardless soldiers, brave in their innocence, strong in obedience and discipline, the common school trains them for the evolutions of life's battle, and its work is nobler than the tactics taught at Woolwich or St. Cyr

---

The banner of the cross was first borne to this coast, more than one hundred years ago, by the devoted padres who came as missionaries to the Indian tribes. The story is full of fascination. Its points are glowing with the national warmth of the Spanish character, suffused with a religious zeal that counted itself happy in the discovery of obstacles and the presence of danger. The story of the Spanish missions has been told many times, but its interest is not exhausted by repetition. All along the coast from San Diego to San Francisco stand the mission churches, many of them more than a century old, their adobe walls defying the abrading blows of time, to which many newer and more pretentious structures have yielded. The mission fathers brought with them wheat, the olive and vine; for bread and wine and oil are the elements in sacrament and ceremony dear to the believer's heart. So it came to pass that the three leading products of our soil, upon which now tens of thousands depend for support, were planted first by holy hands and consecrated to use in the mysteries which are around the lintel of that low door by which we enter immortality.

Following this venerable establishment, as other peoples and other creeds were lured to the new land, came all the communions, and with them to the different pulpits such strong men as are always in the front. The Presbyterian, Baptist, Methodist, Christian, Episcopalian, Congregational, Advent, Unitarian, Universalist and all others, soon floated their standards, and go where you will it is not possible to get beyond the influence, or far from convenient resort to the temple of God that shall best accord with your tastes and convictions. Here the Israelite has built noble synagogues and in them cultured rabbis unroll the scroll of the Pentateuch; and here cathedrals and churches of fine and noble architecture attest at once the piety and liberality of a people who look through nature up to nature's God, with a vision clarified by daily observation of the beauties and the blessings created here and planted in their place by an Almighty Hand.

It was a prudent mother who objected to Ben. Franklin as a husband for her daughter because he was a printer and a newspaper man, and there were already two newspapers in America and she thought the business was so overdone that the cup would never be found in Benjamin's sack.

Since her day the press has wonderfully multiplied, not only in America, but all over the world. A very patient statistician has compiled some interesting figures as to the total number of newspapers and other periodicals published in every part of the world, and brings the total number up to 35,000, thus giving one to every 28,000 inhabitants. Europe, according to these calculations, has 20,000 newspapers, Germany coming first with 5,500, of which 800 are published daily; the oldest being the *Post Zeitung*, published in Frankfort in 1616, while the one with the largest circulation is the *Berliner Tageblatt*, which prints 55,000 copies. Great Britain comes next with 4,000 newspapers, of which 800 are published daily; while France has 4,092, of which 360 only are daily. Italy comes fourth, with 1,400 newspapers, of which 200 are published at Rome, 140 at Milan, 120 at Naples, 94 at Turin, and 70 at Florence, the oldest being the *Gazetta di Genova*, first published in 1797. Twelve hundred newspapers are published in Austro-Hungary, of which 150 are daily, the most remarkable of the Austrian journals being one called *Acta Comparationis Literarum Universarum*, which is a review of comparative literative literature, with contributors in every part of the world, each of whose articles is printed in its native tongue. Spain has about 50 journals, of which a third are political; and Russia has only 800, of which 200 are printed at St. Petersburg and 75 at Moscow. Several of these journals are printed in 3 different languages, and there are also 4 published in French, 3 in German, 2 in Latin and 2 in Hebrew, besides several others in Polish, Finnish, Tartar and Georgian. Greece has upward of 600 newspapers, of which 54 appear at

Athens, while Switzerland has 450 and Holland and Belgium about 300 each. There are 3,000 journals published in Asia, of which no fewer than 2,000 appear in Japan; but in China the only newspapers not published by residents at the treaty ports are the *Ning-Pao*, an official journal published at Pekin; the *Chen-Pao* and the *Hu-Pao*, published at Shanghai, and the Government journal, which was brought out in Corea last year. There are 3 newspapers published in French, Cochin, China, and 1 in Tonquin *(l'Avenir du Tonkin)*, the rest of the newspapers credited to Asia appearing in India, with the exception of 6, which are published in Persia. Africa can boast of only 300 papers, of which 30 appear in Egypt and the remainder in the colonies of England, France, etc. The United States possess about 12,500 periodicals, of which 1,000 are published daily, the oldest being the Boston *News*, which was first published in 1794. Among the United States journals there are no fewer than 120 edited and published by negroes, the oldest of these being the *Elevator*, which was brought out of San Francisco 18 years ago. Canada has 700 newspapers, a considerable proportion of which are published in French; and in South America, the Argentine Republic comes first with 60 newspapers. Australia has 700 journals, nearly all published in English, and the Sandwich Islands 8, of which 5 are in English and 3 in the native tongue. Out of the 35,000 periodicals enumerated above, 16,500 are in English, 7,800 in German, 6,580 in French, 1,600 in Spanish, and 1,450 in Italian.

The oldest newspaper in the New Empire is the *Alta California*, San Francisco. It pioneered the way for a numerous succession. Throughout the States and Territories of the Pacific slope newspapers are thickly planted. In Arizona they plan campaigns against the Apaches. In Utah they skirmish over the Mormon question, and its *pros* and *cons* are served out with great heat. In Nevada the old glories of the bonanza time occupy them with ancient history, while the State's growing agriculture, horticulture and live stock interests, as well as the new mines which keep up its mineral reputation, give the press material themes. In California is a country press of peculiar power and intelligence. It is a faithful reflex of the interests and conditions attractive of immigration, and its unstudied notes of rural matters are a treasury of valuable information.

The metropolitan press is enterprising, as becomes the news medium that hangs upon the edge of a continent, in the Anglo-Saxon commonwealth most remote from the center of that race, London. In a world by itself, an empire within an empire, the press of such a community has functions novel and unknown to journalism in the midst of millions of people, and in vital contact with the dense populations which generate the myriad events we call news. The metropolitan journals of the New Empire get their news over vast spaces of land and sea, and its arrangement, assortment, adaptation, and condensation, call for a tireless industry, and a cosmopolitan intelligence,—knowledge of men and events, and an insight, foresight, and hindsight, that are not required in any other position in the world.

Judged, then, by its schools, its pulpits, and its press, the New Empire may boast that civilization is planted here, and that the temples of learning, and religion and the press, join in guarding the progress of the people in prosperity, and the gentle arts that make up the intellectual pleasures of life, and so add to its enjoyments.

CHAPTER XII.
BENCH AND BAR.

PEOPLE are careful about permanent investment in any country until they know that life, liberty and property are made secure, and have their rights intrenched in an organized judiciary which brings virtue and intelligence to the guardianship of those institutions which mark the difference between civilization and savagery. All of the New Empire had to be wrested from an original proprietorship by methods involving a show of force that does not belong to the judicial arm. For a time after this transfer it was believed throughout the world that there was little security here for what a man carried beyond his own capacity to protect it. But within thirty years of a beginning that was in legal chaos, a sway of law has been established, and in the Territories the Federal Government has planted courts to which men resort for a determination of their rights; and in the States the people have supplied an elective judiciary not inferior to that of the older commonwealths. As business adventure brought here the flower of youthful activity, and here it ripened into business careers the most successful and remarkable that the Republic has seen, so here the best culture of the law schools, and the finest capacities in the legal profession, came. A more brilliant, learned and upright bar has scarcely been seen in America than was the result. The men composing it found here great legal questions, in bold outline, and dealing with them, our lawyers, to a degree, escaped that species of professional controversy which, while it may sharpen, tends to narrow the mind.

Ben. Franklin desired that judges should be chosen by vote of the lawyers, because they would always choose the best lawyer, in order to distribute his practice. By natural choice the people of these States have done this good office for the profession, for almost without exception the judges of all the courts have been selected amongst the ablest and, of course, the most successful practitioners.

The effect of this process has been the rapid spread of the institutions of civilization. Sheep do not feed where the wolves frequent, and property is not accumulated where the laws and the courts deny to it adequate protection. Here the New Empire is happy in courts that stand sternly in defense of all rights, that do not huckster justice, and that form, therefore, no inconsiderable agency in the attraction of capital and the luring of immigration to take advantage of the splendid resources which here await development.

To these courts a president of the United States has come to get a recruit to the ablest side of the Supreme Court of the United States, and for men to fill responsible posts in the Territorial judiciary, where their training fits them admirably for planting and maintaining the forces of society, and laying the judicial foundation of States.

CHAPTER XIII.

MONTEREY.

THE pleasant resorts of the Pacific Coast are outgrowths of the wealth and social taste of the people of the New Empire. There are fine beaches at Santa Monica, San Pedro, Monterey, and Santa Cruz. "Bull Run" Russell, who visited them a few years ago, accompanied by the Duke of Sutherland, declared them to be amongst the finest bathing beaches in the world. They nearly all have passed through the camping stage. Their sands were found to be mellow, and their waters temperate, and camping parties took their tents and leisure there. All that has grown up on shore is simply evolution from the tent and camp-fire. But the air and water are pure as when their advantages were enjoyed *al fresco*. The greatest development has been made at Monterey, on the bay of the same name. Sir Francis Drake, all hero and part corsair, missed Monterey Bay as he sailed up the coast, which he named New Albion, and that placid crescent was not discovered until 1602, by Viscanio. It is noteworthy as having been the scene of the first attempt to take California for the United States, and as the theater of the final affirmation of our title to the soil.

So Monterey is a sort of Plymouth Rock for our Pacific possessions, and therefore the blarney-stone of the New Empire. In 1842 Commodore Jones, of the American navy, sailed into this bay, assaulted and captured Fort Monterey, and ran up the stars and stripes; but soon ran them down again, and apologized. His apology seems to have been accepted, probably because the garrison was short of powder. The incident suggests the former enterprise of our navy, which let no good-looking coast languish for an owner. The Commodore was only four years ahead of time, for July 7, 1846, an American frigate sailed up to Monterey. Her marines did a bit of scuffling with the natives, and the stars and stripes went up to stay. A few days later a British admiral, who was also out hunting land, sailed up to Mon-

CYPRESS POINT, MONTEREY BAY.

terey, to find himself a little too late. The town remained the capital of California for some time, and there met the first Constitutional Convention. To this day it retains many of the quaint Spanish features; and adobe houses, tile roofed, with their ample verandas and high-walled gardens, rouse visions of secluded ladies and sighing swains.

Here, where the sands of the beach are silkiest, the water pure as a maiden's heart, and its embrace warm and wifely, is the Hotel del Monte—"thou most beauteous inn." Around it is a park of hundreds of acres, shaded by the original live-oak trees, re-enforced by magnolias and every kind of great and lesser tree and shrub that the most tasteful landscape gardening requires. Lighting this verdant park, as the constellations do the serene heavens, are acres of flowers; and through it, sinuous and graceful, wind drives and paths, tempting to lovers.

In the midst of this bloom and perfume stands the hotel, the perfection of adaptation to the conditions of a seaside resort. Fire-places cheer the evening, for remember that the waters of Del Monte tempt the surf bathers in what the East knows as the winter months, and night-fall makes the open fire a feature amongst the comforts of life. But here no winter; the flowers bloom, the trees flaunt their green ers, and in the open waters of the great bay the whales s in all months. In the Del Monte are spacious billiard rooms and there you may play ancient "shuffle board," which was the diversion of Shakespeare and Sir Walter Raleigh, before caroms and cues had been invented. "We have had pastimes here, and pleasant games."

It is impossible to refer to the material resources and latent wealth of the New Empire without at least this much reference to a resort created by its fashion and good taste, since civilization points to these refinements as proof of its existence, and life is softened by occasional indulgence in the recreations to which they tempt. It is well that Del Monte is planted to face the waters that first fell under our dominion, sheltered by the mountains that gave back a replication of broadsides, whose iron voices proclaimed conquest and decreed liberty to this land. In it are compact the graces and gifts of Saratoga, Long Branch, and Cape May, and, in the judgment of men and women of the best taste, it is one of the most pleasure-provoking resorts on either coast of the continent. Like the growth of our cities, it is a sign of the enterprise of our people.

CHAPTER XIV.

MODES OF TRAVEL.

THE New Empire has become the highway of the world. It has realized Benton's dream of a new path to India, reaching the East by going west. Columbus had that plan in view when he sailed west, and ran his prow into a continent that lay across his track, stretched almost from pole to pole. At first this continent was an obstruction to travel, but it has been turned into a facility. In Benton's speech to the Senate he proposed to build a monument on the summit of a Rocky Mountain pass, with a hand pointing to sunset, and inscribed upon it, "It is the East! It is India!" So across this continent travel was at right angles to the meridians of longitude, while its great rivers paralleled them. Inland navigation was mainly north and south, following the rivers. Travel, following the instinct which led our race out of western Asia, and set its face westward, and has brought it to the edge of this continent, facing its birthplace, could not use the rivers, so it crossed them.

No people ever came through greater difficulties than those met by the early Californians. They had choice of three routes, around Cape Horn, overland by wagons and pack trains, and across the Isthmus. Around the Horn required six months, and exposure to every extreme of climate; for on the Atlantic it was a plunge from the north temperate zone clear through the tropics, across the south temperate, into the south frigid, and a repetition of the same experience in the Pacific with the order reversed. Many a man fell under the perils of the long voyage, and many a ship laid her bones around the stormy cape or in the Straits of Magellan. The necessities of so long a voyage put the passenger upon rations of salt food and bad water, and often scurvy rotted the flesh on his bones before he reached a diet that could arrest its ravages. That voyage of a half year, through tropical storms and polar snows, with hardship and disease hovering every knot made by the ship, was so full of discomfort and so often fatal to those unaccustomed to going down to the sea in ships, that its alternative by the Isthmus came into favor, and the tide of travel turned toward Panama. Any hull that would hold an engine became a steamship, and the reeking Isthmus was traversed in canoes and flat boats and rafts as far as its rivers were open to such transit, and then mules and horses were substituted. Cholera and yellow fever lurked by the way-side, and struck down many a strong man with the suddenness of a thunderbolt, and many a youth there gave up his life and with it the golden hope that had lured him into this lair of death's twin furies. At last the Panama railroad was built by an outlay of life that made every tie represent the bones of a laborer, and over this highway, digged by death and bordered by an unbroken line of rattling skeletons, poured the tide of life. The third method was overland. Under favorable circumstances the journey could be made in six months, from the Missouri River, along whose banks were the outfitting points, the Irak, Damascus, and Cairo of those more earnest than Meccan pilgrims. The line of civilization, held by the advance guard, then lay east of that river; west of it the geography and geology were as nebulous as to-day they are in the interior of Africa. Across the map was stretched a blank space, usually colored yellow to make a meaner impression, and named the Great American Desert. The desert began within fifty miles of the west bank of the Missouri, and its Nile was the Platte, which came boiling down, roily and treacherous, useless for navigation, and hard to ford, for its quicksands were always hungry and had their fill on many a stout ox-team and band of horses. The forty-niners had to face the imaginary and real terrors of a trip which, alas! is no more a necessity. It lay through the territory of wild Indians, who levied tribute on the wagon trains; who stood at fords and ferries and exacted a toll that now pays a fare across the continent from ocean to ocean, with six days instead of six months required for the trip. It sometimes seems a pity that the terrors and toils of these three primitive routes to this coast should be no longer, for they were a magnificent test of the endurance and courage of men; but they made martyrs, and each of the early grand highways has its tale of death and suffering. The fancy will never tire of the story of Herndon preferring to sink with his ship, nor of the tragedy of Mountain Meadows so tardily avenged, nor of the snow and famine that closed around the Donner party and imprisoned them to starvation. Along that overland way gentle women were brought to the agonies of maternity by the camp-fire, and to young and old came the final summons which must be obeyed equally in the desert or the city.

But what a change was wrought within nineteen years of the beginning of immigration to this coast! From being the least accessible, the hardest to reach, and most difficult to leave of any part of the Union, by the completion of the Union and Central Pacific Railways it became easily accessible. The terrors of Cape Horn, the fevers of the Isthmus, the perils of death by thirst and famine overland, all passed like the morning gloaming. Now, four routes by rail, the Central Pacific, Southern Pacific, Northern Pacific, and Atlantic and Pacific, connect the New Empire with the East; and that land so lately reached through perils that would have appalled a Crusader, is brought within six days of the Atlantic Coast, and the citizen of London who has business in Melbourne or Calcutta takes his through ticket *via* San Francisco, finds his berth in the sleeper waiting him at New York, Omaha, and Ogden, and in fourteen days from Liverpool sees the tide come through the Golden Gate. And the fare for the trip costs less than the price of a single team to be used in the long overland journey of thirty years ago.

We take the goods the gods send us, as a matter of course due to our deserts. But who shall estimate the business toil, the readiness, and adventure which led railways up mountains and down, through wastes where no dewdrop catches the sun rise, overcoming snows and shifting sands, and so made the whole world that goes from the Occident to the Orient, pass by the Golden Gate! What a change this enterprise wrought! No more double transit of the equator, no more deathly wrestle with yellow fever, no more whoa-hawing of the patient ox overland, but, instead, six days of luxurious travel, on quintuplex springs and paper wheels, with a bed at night full length, and a companionship pleasant, because sure to be a miniature congress of the nations.

CHAPTER XV.

THE WORLD'S FAIR.

GRAND expositions of the industries and productions of the earth, its inventions, machinery and art, illustrations of the state and progress of science, have been held in London, New York, Paris, Philadelphia, Vienna, and New Orleans. None has been so located as to collect the resources and immediately interest the peoples of the great countries bordering the Pacific. The next exposition should be located in San Francisco. It is in the center of the territory of the United States, and is the greatest seaport on the Pacific Ocean. It has ample railroad connections with Mexico, will soon have with British Columbia, far Manitoba and the Hudson's Bay country, so that its railroad facilities will outreach from the tropics far toward the polar circle, while with interior and Eastern and Southern States they are ample. Since the New Orleans Exposition it has been said that another cannot be undertaken for a very long time, because of the discouragement its failure has caused. There is nothing in this. The great cities of the world are located in a belt that lies between 38 and 51 degrees north latitude. Within that zone are located the great activities of the human race. Within it are the industry, thrift, economy, and enterprise which have generated the capital that controls the productions of the globe. The commerce and travel that are in ceaseless motion are confined to that circle clear around the earth, for within it are London, Paris, Berlin, Vienna, Madrid, Rome, Constantinople, Pekin, Tokio, San Francisco, Chicago, St. Louis, Baltimore, Philadelphia, New York, and Boston, with a city population aggregating nearly twenty millions of people. It is the world's commercial zone, and temporary concentration of its activities and their results has always been easy within its borders, but never a success outside.

San Francisco is one of the jewels set in this ring around the earth, and here are all the natural advantages and features which tend to make a successful world's fair. The great bay, the mountains that border it completely around, the natural objects of interest within easy reach, the geysers, petrified forests, mineral springs, forests of great trees, the mountain peaks that are easily climbed above the clouds, the valleys covered with vineyards and orchards, the hills clad with bright olive groves, the orange orchards flecked with golden fruit and aromatic with bloom, Yosemite Valley, Lake Tahoe, and the sunny sea beaches, make up a combination that cannot be equaled by any other locality in the world. The cities of the bay, San Francisco, Oakland, Alameda, Berkeley, San Jose, San Rafael, Saucelito, Vallejo, Benicia, and Napa, all within a short reach of each other either by rail or steamer, offer ample accommodations to a congress of the nations. The cosmopolitan nature of our population is an attraction that no other city in the world can offer. Here are settled Greeks, who can welcome their countrymen from the Ægean; the Bombay merchant will find here his Parsee brethren; the Japanese will hail his friends, and the Chinaman will find here the many buttoned mandarin of the Flowery Kingdom. Our German, French, Italian, Spanish, Portuguese, and Russian fellow-citizens are in such numbers as to be appreciable in the social life and business of the New Empire, and to materially influence the favor of their countrymen towards a world's fair in San Francisco. The hotels here are on a scale of amplitude found nowhere else, and they stand at the head of the hotels of the world, unexcelled by any in architectural effects, capacity, and administration.

On what better or more accessible ground can the world's captains of industry summon a general muster than this? It is within the commercial and industrial belt, and right in the path of circum-terrestrial travel. New Orleans was at one side, inconvenient of access, and unfitted for such an enterprise. Asia, Africa, Australia, and the islands of the sea will vie with Europe and America in showing here the highest results of the toil, genius, and art of their people, and so there will come to hundreds of thousands an opportunity to see the New Empire and at the same time show what the old empires have done and are doing for the advancement of mankind.

The world has never tested fully the hospitality of the people of the New Empire. Here are scores of the grandest and roomiest private houses on the continent, with owners whose keenest pleasure lies in the generosities of entertainment, and to which even visitors with crowned heads may resort, to confess that they have never enjoyed more the pleasure of being guests.

But aside from this, here is the cheapest living in America, and the best, with a market that never fails in the choicest meats, fish, poultry, vegetables, and fruits anywhere grown. Here is associated the greatest economic skill in its preparation for food, so that the restaurant fare of San Francisco has come to be noted all over the world for its excellence and cheapness. The advanced guard of visitors to a world's fair here will have no tales of bad service and extortion to send back to deter others, as was the case at New Orleans; but, rather, the skirmish line will ask the main body to come on, for here they can enjoy all the comforts and luxuries of life as cheaply as at home.

CHAPTER XVI.

CALIFORNIA.

THE population of California is given at 1,000,000, which is being increased by births and immigration at the rate of 60,000 per annum. California, with her resources properly developed, is capable of sustaining a population of 20,000,000. The assessed value of her real estate foots up $500,000,000; personal property $200,000,000; 7,000,000 acres of land are under cultivation, and 9,000,000 acres are fenced. The value of annual products is $180,000,000. As a State, she is practically out of debt. In her savings banks are deposited $60,000,000. The banking capital of the State is $50,000,000, and the annual product of bullion is $18,000,000. The average value of the wheat crop is $45,000,000; barley, $10,000,000; dairy products, $8,000,000; fruit crop, $7,500,000; wool clip, $8,000,000; wine products, $5,000,000; value of lumber manufactured in the State, $5,500,000; hay crop, $13,000,000; domestic animals of all kinds, value, $60,000,000; value of animals, poultry, etc., slaughtered every year, $23,000,000; increased value imparted to manufactures, etc., by labor, $40,000,000; number of grape vines set out, 130,000,000; fruit and nut trees, 800,000; with five times as many forest, shade, and ornamental trees. The State contains 3,500 miles of telegraph lines, 3,300 miles of railroad, 5,000 miles of mining with an equal extent of irrigating ditches; 400 quartz mills, 300 saw-mills, and 185 flouring mills; $250,000,000 have been invested in mining improvements in the State, cost of quartz mills, tunnels, and ditches included.

The annual reports of the Agricultural Department at Washington, running over a period of three years, with a general average, show the following interesting facts with regard to some of the productions of California, as compared with all the other States and Territories in the Union, and more particularly with regard to some of the leading agricultural States:—

BARLEY.

	BUSHELS
Entire production of the United States and Territories, including California	42,564,692
California	14,723,915

HAY.

Average value per acre, United States	$12 34
California	21 65

OATS.

Average value per acre, United States	$ 8 84
California	20 55

POTATO CROP.

Average value per acre,	United States and Territories		$42 74
"	"	California	97 29
"	"	Oregon	62 57
"	"	Kansas	51 99
"	"	Michigan	45 42
"	"	Minnesota	36 54
"	"	Nebraska	33 33
"	"	Illinois	36 18
"	"	Iowa	29 56
"	"	Wisconsin	32 74

AVERAGE CASH VALUE PER ACRE OF CHIEF AGRICULTURAL PRODUCTS.

United States and Territories, including California	$12 68
California	18 36
Texas	15 68
Wisconsin	12 34
Illinois	10 59
Minnesota	9 86
Iowa	8 91
Florida	8 64
Kansas	8 22
Nebraska	7 34

CORN CROP.

	VALUE PER ACRE.	YIELD PER ACRE.
United States and Territories, including California	$10 13	27.90 bu'ls.
California	22 38	31.50 "
Colorado	19 75	26.20 "
Iowa	8 33	37.80 "
Illinois	8 80	29.80 "
Kansas	7 95	32.10 "
Florida	7 15	8.09 "
Nebraska	7 60	38.00 "
Georgia	6 52	9.80 "

And also in the value of her wheat crop California leads all the States and Territories in the Union. Ohio stands next, Indiana third, with Michigan, Minnesota, and Illinois following in succession.

The following table of wheat and barley acreage for 1886 has been carefully compiled from reports just received from correspondents in the principal grain-growing counties of the State. It is of significant importance, as showing not only the acreage in wheat and barley for the present year and the average yield of each per acre, but also that the wheat crop of California for 1886 will be much larger than the greatest wheat yield during any year of any State in the Union:—

County.	Wheat Acreage.	Avr'ge Y'ld per Acre. Centals.	Barley Acreage.	Avr'ge Y'ld per Acre. Centals.
Alameda	75,062	10	32,373	15
Calaveras	30,000	8½	12,000	12
Colusa	400,000	11	45,000	16
Contra Costa	153,360	11	29,040	15
Fresno	300,000	9	40,000	15
Kern	15,000	11	8,000	18
Los Angeles	250,000	11	100,000	20
Mariposa	1,500	11	6,700	12
Mendocino	12,000	11	4,000	17
Merced	185,000	8½	27,000	22½
Napa	25,593	10½	3,211	13.70
Sacramento	87,000	10 1-5	22,000	13 1-5
San Benito	47,000	13½	9,000	24
San Bernardino	3,500	11	80,000	19½
San Joaquin	250,000	11	40,000	21
San Luis Obispo	101,000	11	125,000	15
San Mateo	8,000	13	7,000	20
Santa Barbara	60,000	13	20,000	20
Santa Clara	101,355	14	168,935	22
Santa Cruz	25,000	14	11,000	20
Siskiyou	10,000	10	6,100	14
Solano	61,536	9½	16,770	13.24
Stanislaus	341,000	9	45,000	10
Sutter	95,000	13	18,000	17
Tulare	415,000	10	45,000	15
Tuolumne	7,025	10	2,580	15
Ventura	20,000	9	80,000	13
Yolo	340,000	10	65,000	15
Yuba	30,000	10	13,000	20
Totals	3,450,131		1,081,729	

All the counties of the State are not here enumerated, because the assessors had not received enough sufficiently accurate returns on which to base reliable statements, but general reports received from these counties show that both the average and yield of them will be in excess of last year.

The total acreage sown to barley as shown in the above table is 1,081,729 acres. The total yield of barley from these counties as calculated out is 18,633,130 centals, equal to 38,819,020 bushels. The barley crop might have shown a still larger return, but in many counties large quantities were cut for hay, which, had it been allowed to mature, would have made good marketable grain.

The total acreage of wheat as shown in the counties mentioned in the above table is 3,450,131 acres. The total yield as figured out is 35,862,518 centals, equal to 59,770,863 bushels.

There are, as above indicated, still a few counties to hear from which, it is fair to assume, will enlarge the production so that the wheat crop of California for this present year will reach the enormous quantity of 60,000,000 bushels.

Asbestos is found in many counties of the State, and is mainly utilized as a coating for steam boilers and pipes.

Ores of nickel occur here also, but not in quantities sufficient to be profitable to work.

The only extensive deposits of chrome ore in the United States are found in this State. They are mainly found in Placer, San Luis Obispo, Del Norte, and Alameda Counties. About 3,000 tons per annum are shipped to Baltimore and Philadelphia.

The joint production of borax of California and Nevada has increased from 5,180,810 pounds in 1876 to over 8,000,000 pounds in 1885. The Pacific Coast exports in 1885 amounted to 9,000,000 pounds. The borax fields are in the boundaries of the two States, and are the only ones in the United States.

California produces about 200 tons of carbonate of soda per annum. It costs $45 per ton delivered in San Francisco.

The Inyo County marble deposit is a very large one, and is now being worked.

California is prolific in limestone, there being several extensive belts. Some 220,000 barrels of lime were manufactured in the State in 1885.

Several deposits of manganese exist in California, but only one or two are being worked.

Very little cement is made in the State, although there are deposits of hydraulic limestone, and there are two cement factories, one at Benicia, and one at Santa Cruz.

The manufacture of plaster from the California gypsum deposits has increased of late years. Some 2,500 tons of gypsum were ground by the mills in San Francisco in 1885.

Petroleum is found in Humboldt, Colusa, Contra Costa, Alameda, Santa Clara, San Mateo, San Benito, Ventura and

Los Angeles Counties in California. The product of the State has increased from 15,000 barrels in 1878 to 325,000 barrels in 1885.

Antimony occurs in several places in California, San Benito and Kern Counties each possessing producing mines.

There are several quarries of building stones, some of which are being worked.

Large quantities of salt are consumed on the Pacific Coast, much being needed by reduction works. In 1885 31,000 tons of salt were made in California, mainly in Alameda County, on the shores of San Francisco Bay.

A great deal of asphaltum is mined in the State, and is utilized at home.

The clays are utilized by the potteries in various parts of the State, mainly in making the lower grades of pottery.

The State produces about 1,200 tons of metallic copper per annum.

Graphite occurs in many localities, and some few of the deposits are utilized.

There is only one iron mine in the State that has been worked, in Placer County, but low prices in 1886 have caused the furnaces to be closed down.

Among other mineral products of the State are alum, bismuth, iridium, platinum, lithographic stone, mica, and sulphur.

SCHEDULE OF RATES OF WAGES PAID.

	PER DAY.	
Carpenters	$3 50	
Machinists	3 25	
Sign-painters	4 00	
Boiler-maker	3 50	
Tin-smiths	3 50	
Longshoremen	3 50	
Stone and marble cutters	4 00	
Plasterers	4 50	
Gun and locksmiths	3 50	
Roustabouts	2 50	
Coal miners (shift work)	2 50	
Coal miners (by the yard)	3 00 to	4 50
Mechanical Engineers	3 00 to	4 00
Bricklayers	5 00	
House painters	3 25	
Pattern-makers	3 50	
Shoemakers	3 00	
Blacksmiths	3 50	
Day laborers	2 00	
Gas-fitters	3 50	

Upholsterers	$ 3 50
Boat builders	3 50
Plumbers	4 00

PER MONTH.

Tailors	$54 00
Mill hands	60 00
Bakers	60 00
Farm laborers (with board)	30 00 to 40 00
Loggers:—	
Teamsters	40 00 to 65 00
Choppers	65 00 to 70 00
Skidders and hook-tenders	55 00 to 60 00
Swampers	50 00
Sawyers	50 00 to 55 00
Common laborers	40 00 to 45 00
Boys	30 00
Cooks	50 00

THE PRESENT CHINESE POPULATION OF THE STATE AS SHOWN BY THE ANNEXED TABLE:—

(Compiled from Returns made by the County Assessors and County Clerks.)

County.	Population.	County.	Population.
Alameda	8,000	San Bernardino	371
Alpine	125	San Diego	275
Amador	2,000	San Francisco	43,000
Butte	3,000	San Joaquin	2,500
Calaveras	1,037	San Luis Obispo	196
Colusa	1,500	San Mateo	250
Contra Costa	500	Santa Barbara	500
Del Norte	300	Santa Clara	2,950
El Dorado	400	Santa Cruz	450
Fresno	753	Shasta	1,335
Humboldt	250	Sierra	400
Inyo	50	Siskiyou	1,458
Kern	702	Solano	995
Lake	469	Sonoma	1,500
Lassen	50	Stanislaus	700
Los Angeles	5,000	Sutter	550
Marin	350	Tehama	774
Mariposa	600	Trinity	400
Mendocino	1,000	Tulare	1,000
Merced	550	Tuolumne	500
Modoc	60	Ventura	300
Mono	363	Yolo	400
Monterey	500	Yuba	2,000
Napa	650		
Nevada	1,800	Total	90,022
Placer	2,190	Total from tenth census	8,618
Plumas	500		
Sacramento	3,000	Grand total	98,640
San Benito	87		

But the Chinamen are rapidly leaving the State, and so not only making room but creating a demand for good white labor. There is no country in the world where honest toil is more handsomely rewarded in proportion to the cost of living than in California.

Society here is as well organized, and devoted to the good works which are the merit of a great people, as anywhere in the Union.

The climate is so full of blandishments that it tends to attract the best population from all parts of this country and Europe; and its guarantee of good health, and the enjoyment of life which it permits, its tendency to development, activity, and refinement, its decided effect upon the literary and artistic character, which it develops to a wonderful degree, will focus here the growth of art and science.

The mineral and thermal springs of California, with established curative powers, and in situations unequaled in the romantic interest of their scenery, will one day outrival the great spas of Europe, to which so many sick make long pilgrimages.

All of the financial, insurance, manufacturing, commercial, and rural interests and industries are here in the hands of the country's best enterprise and intelligence.

San Francisco is the center of a greater whaling industry than New Bedford. It has the largest trade in peltries of fur-bearing animals and amphibia in the world.

Ostrich farming is being rapidly transferred from South Africa to Southern California, where it is demonstrated to be a most profitable success.

We will soon rival France and the Mediterranean slopes in our wine and oil trade, and our mineral interests in gold and silver will long lead the world, as now. We ship thousands of car loads of oranges, lemons, and limes, and the citrus orchards are every year extended.

In fine, no matter what a man's tastes and fancy as to occupation, here he will find a country of opportunities, amongst which he is sure to make an agreeable choice; and his selection made, he will find in it full and happy scope for his most wholesome energies, with the certainty of more adequate reward for the efforts he invests, than any other country can offer. So it is that thousands have looked down upon this promised land, as they approached it, poor in all things but hope and industry, who are now affluent, and the stewards of a heritage of comfort for their children.

CHAPTER XVII.
IRRIGATION.

SINCE the prospectus of this book was issued and sent East to those centers of population resorted to for information by intending immigrants, many inquiries have come back in relation to the certainties of irrigation; and since the recent adverse decision of the Supreme Court of California, these inquiries have taken a discouraged tone, which we desire to correct by a statement of facts.

The wonderful growth and prosperity of Southern California are associated in the Eastern mind with irrigation, and properly so. The visitor who now revels in the luxury that is found at Pasadena, in the San Gabriel Valley, at Riverside, Santa Ana, Anaheim, and scores of places in that section of the State, has only to note the parts of plain and mesa yet in their natural state to see the magic wrought by water used for irrigation. Witness, too, the marvelous results conjured out of the deserts in Kern County, where irrigation has spread green fields of alfalfa and yellow fields of grain; where orchards, vineyards, fields of cotton, and the cattle fat as those that ran on a thousand hills, have taken the place of desolation, once the home of the serpent, the centipede, and tarantula.

Here is the same sharp contrast so often noticed in Los Angeles, San Bernardino, and San Diego Counties. The Kern irrigating system has hundreds of miles of canals and ditches, and the zones they irrigate are perfect pictures of plenty and prosperity, scented with the perfume of flowers,—an Arcadia, where labor is light as the leisure of life elsewhere. On the higher zone, above the ditch or canal, is the desert, thirsty, gaping, unpeopled. It is the rough diamond, with its beauties undeveloped; while, where the water has worked its ministry of regeneration, it is the diamond fresh from the lapidary's wheel, a thing of beauty and a joy forever.

There be those who have explored these wonders of Kern, who have noted how, by seeking still higher levels for tapping the streams, zone after zone may be reached by the waters of life, who call it the foremost county in California in natural resources and capacity to support a dense population. The Eastern visitor should resort to it as an extreme illustration of what irrigation can do, and then should consider that if water can produce the charms, the profits, progress, and prosperity spread abroad there, where even a cony could not live before, what may irrigation not do extended to the whole arid area of the San Joaquin and Sacramento Valleys?

In this projected inquiry the investigator is not left without illustrations. He will find them in the Fresno colonies. These are oases created by irrigation. As a rule they are divided at right angles into twenty-acre tracts, with main avenues lined by palm, shade, and nut trees.

These twenty-acre holdings are visions of the beauty of high farming. They have not been settled by the rich, but by the poor in purse. It is one of California's most promising raisin-producing areas. The climate is warm, dry, and bracing; and the soil, when coaxed by water to surrender its treasures, proves the most generous in the world. We have before us the history of the Fresno colonies. After the vines planted on them reach three years of age, the minimum return, net per acre, is $100 per year, and the tillage and care of the crop is within reach of the members of the family. Result: Those colonists who settled there poor, are now in comfortable circumstances. To illustrate: Nine years ago a Swede immigrant named Anderson landed there with a wife and seven children, and $75.00 in cash. He took one of these small tracts on credit; worked for wages while he improved it, and, at the end of nine years, sold his land for $12,000, having meantime got out of debt and supported his family comfortably. Talk about no chance for anybody in California! Irrigation offers not chances, but certainties to a denser population than any of the older States can boast. Another Fresno case: Miss Austin, an Oakland schoolteacher, with health weakened by her arduous profession, and only $1,000 in her pocket, went to a Fresno colony eight years ago and bought twenty acres. She had it planted in vines, and while waiting for them to bear, bought grapes of her neighbors, dried and packed them in neat packages, which gave them a special character in the market. To-day she owns forty acres, all improved; is out of debt, and it is safe to rate her worth $50,000. So I might cite case after case. Can they be equaled in the rural records of any of the great agricultural States?

Now what has done it? As we have explained in our chapter on topography, these plains and deserts have had washed down to them the richness of the mountains. The mountains are the mighty bones of the continent, and, cracked by past volcanoes and present beating storms, their marrow has run out for enrichment of the plains. All the streams have their final source in the mountains, and their waters continue to bring down this marrow, so that when irrigation puts them on the land, and it begins producing crops, it is not exhausted, for every irrigation is a process of re-fertilization of the soil. Here is no need to buy phosphates and guano. The dung fork is an unknown farm tool. Instead of buying a cart to haul a manure heap onto the land, the Fresno farmer can put his cash into a phaeton for his wife. Here, then, invoked by irrigation, is the ideal rural life. No prayer goes up for rain, and an overruling Providence, unannoyed by being continually asked for a drink of water, showers unasked a thousand gifts and graces upon a people who make their own rain, and measure its fall upon the ground. Irrigation was inherited from the Spanish and Mexican owners of the soil. It was recognized by Federal law when the United States owned the streams and the land. When by Government patent they passed to private ownership and State jurisdiction, the right to useful appropriation passed too, and was for years undisputed. During those years there sprang into being all these impressive results which we have hastily sketched. No man believed otherwise than that his use of water for irrigation, being in line with

Federal policy, was in line with local law. Occasionally the right of riparian owners to prevent diversion from the bed of the stream was mooted, and the shadow of the English common law was conjured for a temporary scare. But the irrigators knew that the English common law of riparian monopoly of the waters had been especially and specifically nullified in every English colony and country controlled by that empire, whose physical features and necessities are like those of California. In Australia, and all her Australasian possessions, in India and in our neighbor, British Columbia, an assertion of this moist country law of riparian rights would lose its standing in the courts in a moment. Knowing this, the people of California were fearless in their appropriation of water until, in a legal contest between men who, by costly and beneficent hydraulic systems, had made an Eden where had been a desert; and other men who claimed the right, because their lands abut the stream, to compel its waters to go and waste themselves in the sea, the Supreme Court of California went into the rusty locked closet of "precedent," brought out the fleshless skeleton of English riparian law, fit emblem of famine, and have tried to wave its bony hand over the orchards and gardens and vineyards, to wither vine and olive tree, and blight the grain and fruit. The result has been the most powerful, spontaneous, popular movement ever seen in this Union. It is more general than that rush to arms when civil war was upon the land. In all the great cities and in every rural community the people are banded in organizations.

A powerfully representative State Convention held in San Francisco has crystallized these aroused energies, and guided them to aim a solid blow, whose impact no Court can withstand. The grievous decision of this bench was reached by a majority of only one vote, and in the coming election, without any break in the ordinary procession of events, this majority will sink to a minority. But the forces that are abroad are stronger than they need be to do only this. It is different from any other judicial issue. It is believed by the ablest publicists in the State, and those most respectful to the Courts and their authority (and this belief has found a positive voice in the unanimous press) that the Court should hand its commissions back to the people from whom they were derived, in order that a full bench may come fresh from the masses to reflect the mightiest interests of the State, and entrench them in the law. Before the power now invoked and active in every county in the State, no law imported from abroad, to curse our people with blight and famine, can stand. The measures proposed involve amendments to the Constitution which will grind the grinning skeleton of English common law to powder between the upper and nether millstone of the public will; and they also include statutes which will protect, regulate, and affirm permanently the appropriation of water for irrigation. In all this there is to be no delay. The political party that stands in the way will get run over. The public man who opposes will wonder what hurt him. So we say, to those Eastern readers who have talked of going to Colorado or Utah because there irrigation is a settled policy, and offers a chance for capital to be safely invested in hydraulic works, and assures to land owners the certainty of water for their fields, "Don't be hasty; just bide a wee and witness the speedy and complete adjustment of this California issue in line with the needs of our commonwealth and people, and then come here."

Following will be results which stagger prophecy. The water and the land will go together, and, as the limits of the present volume of water are reached, there will appear a system of storage of flood waters. The contributing cañons in the mountains will be dammed, as is done now on all the rivers of New England to store water for manufacturing power, and as the Federal Government has done at the heads of the Mississippi to impound water in the spring, that the country below may be saved from floods, and that the river may be replenished at midsummer to hold it up to navigable stage. The field opened out here is illimitable.

In this work hydraulic engineers will find employment, and in the construction of canals and ditches and dams, thousands of laborers will find remunerative work. All the industrial energies of the State will feel the impulse of this mighty policy, and the cities will derive from its effects a commerce that will spread their borders and stimulate their business to a prosperity unknown by the cities of the East. Here, for generations, will be the progressive expansion in real estate values and in the margins of business. The trade that will be inspired from this source will add value to our mines and timber, to our manufactures, our fisheries, and to every activity and investment which go to make up the complex industries of a great people. The history of mankind is that the highest primitive civilization was in rainless countries capable of irrigation. This is because men there were relieved from the eccentricities of the seasons, and the produce of the soil, which is the foundation of everything, was made certain. We speak of our present civilization. Its remote source was Egypt, irrigated by the Nile. Greece took her culture from the Egyptians and passed it on to the Romans, and they gave it to all Southern Europe, and its line of march was continually along the zones where the soil yielded its best gifts only when subjected to irrigation. With that civilization so descended art and science have pitched their moving tents; literature and history have told its story as they moved in its van, and poetry has strung its harp and sung of love and war, from the time Miriam chanted her hymn of adoration upon the entry of Israel into the irrigated valleys of the Promised Land, and David wrote in stately measure: "The Lord is my Shepherd; I shall not want. He maketh me to lie down in green pastures; he leadeth me beside the still waters. He restoreth my soul."

CHAPTER XVIII.

IRRIGATION—CONTINUED.

WE have devoted so much attention to the mechanics of irrigation, because useful appropriation of water in California and, to an important degree, throughout the New Empire, is at the foundation of the two great industries which germinate the wealth of all this region. It is difficult to imagine a commerce not derived from our mines or our agriculture, and unless water can be appropriated for the use of both, their profitable pursuit will be confined within limits so narrow that it is idle to talk of them as the foundation of a great commonwealth.

Without the profit of these occupations, the timber of our forests and the fish of our waters, which are the sole remaining means of production, will not be worth the effort it costs to put them into commerce, and it requires no argument to demonstrate that with production, limited or suspended, transmutation and exchange either cease or shrink below a return tempting to the enterprise of men, and so our manufactures, which are now greatening to the demands of a dense population, and have enlisted capital more upon hope and future promise than present profit, will decay, and their machinery will cease its pulsation, and their fires their glow, while the capital that has created the manufacturing plant will be as completely lost as if it were in a grist mill on a water-power which the channel has deserted and left with no water to turn its turbine and buhrs. These reflections go to the philosophical radix of a country's prosperity. A community thrives by making the best and wisest use of its natural facilities and advantages. Amongst the people of Europe the Swiss stand as a type of cheerfulness and patriotism. We seldom hear of extreme poverty amongst them, and they maintain in their simple forms of government the primordial principles which are the ultimate base of our own laws and institutions, as they are in some respects the model of all free societies. The Swiss are not an accident, nor are their manners and customs those of a people by chance light-hearted and prosperous. They have for centuries made the most of the natural advantages furnished by their country, though these are few in number and parsimonious in degree. They have carved their Alpine woods into toys of cunning form, and their handicraftsmen, inheriting generations of skill, have worked wonders in metals. The bits of grass that spring under the drip of glaciers have been treasured to pasture cows and goats, and the *chalets* of that land are the shelter of happy thrift, and her people illustrate the pleasures of contentment.

We have endeavored faithfully to describe the natural resources and abounding advantages of the New Empire, and while cold type must fail to adequately portray them all, we have shown them to a degree which throws in high relief the generosity of nature. Parsimonious to other lands, here she has lavished her gifts with a prodigal hand.

Of those to whom much has been given, much is required, and the decree cannot be escaped by our people. As a measure of their duty in the great question which concerns the availability of all these natural benefactions, we come now to consider briefly the legal aspects of the question of irrigation. We have shown the richness of the land and the abundance of water within the banks of streams. Our California people are now summoned seriously to decide whether this State shall be a thirsty Tantalus, sunk to the chin in waters she is forbidden to drink, or whether law and nature shall be in harmony, and of the abundance she shall slake to her full satisfaction. We hear a deal of the common law. What is the common law? It is the law of custom. What determines custom? It is shaped out of the physical surroundings and natural necessities of a people. Reduced to its simples, custom, the natural common law, dictated by physical necessity, makes an Esquimau dress in furs, sleep in a bag of eider duck skin, live on walrus blubber, and build a house as far as possible impervious to the nip and gnawing of an Arctic winter; while custom, the common law of physical necessity, makes the native of the tropics swing in a hammock, eat bananas, and compromise with decency in the lightness of his dress. Impose upon the Esquimau the habits of the tropics and he would be a frozen monument of folly or despotism in a half hour. Force the diet and dress of the intra-polar circle upon the native of the tropics and he would die loathsomely of surfeit and fever in less than a week. If this make plain the common law of custom, and that is the only common law of any country, let us suppose

California to be a Crusoe's Island, the real *terra caliente* which the old Spaniards supposed it to be when, in default of inconvenient exploration, they called it an island; suppose this land to be settled by people who have upon them the duty of founding institutions, devising a polity and developing a jurisprudence in harmony with the physical conditions which set the bounds to those activities by whose practice they must support life; what would they do? Does any one who knows the chemistry of our soils, the topography and climatology of this region, believe that any one would dare propose that the riparian owner at the mouth of a stream should dominate its waters clear to the mountain rills whose filaments join to make the volume of its flood, and should have the right to forbid that its quantity should be diminished or its quality deteriorated, to the least degree, but that it must all flow wastefully past the borders of his holding? In the primitive society which we have supposed to exist, there would be instant revolt at such a proposition, and the principle behind it would be held a petty treason to the inchoate commonwealth, and why? Obviously because the concensus of horse sense in the community would instantly discern the inharmony between such a rule of riparian regulation and the law of natural necessity, the voluntary custom, compelled by the physical conditions under which these people must live, and from which there is no escape until the god of bounds grows weary of watchfulness, and natural law is lost in a convulsion which issues in chaos. To force such a rule upon a people, situated as those of California are, is like the exchange of customs between the tropics and the Arctic circle. Supposing this people to be free agents, we would find them devising just regulations by which water and land should be brought together, to secure that certainty in returns of rural industry which is the strongest incentive to labor, since it is true that the arm of the sower is strengthened by the certainty that he is to reap; and a tree is planted, watched, and tended with more refined care, when the laborer who does it knows that its fruits are for the pleasure of his own palate, and not so remote in their coming that they are to be enjoyed by another. So this people would write first in their statutes that the use of water should fix the right to it, and that each user should have no right to a drop beyond what he needed, and that with cessation of use his right ceased, and became subject to appropriation by another for devotion to a useful purpose. Out of this customary law, this evolution from natural necessities, would spring a system of rules and regulations framed in regard to the rights of all, and so shaped as to make every drop of water useful upon every acre of land upon which hydraulic engineering could carry it, and this whole system of rules and regulations would be the common law of California upon the matter of useful appropriation of water. We are sure that the reason and reasonableness of this need no further demonstration. Having shown how the customary law would have naturally developed straightly along the line of the right of useful appropriation of water, let us exchange hypothesis for history, and see what was done by the people who first assumed the duty of founding this commonwealth. The pioneer laborer here, after the conquest, was the miner. He at once became an appropriator of water, without which the pursuit of his calling was impossible. Mining in gulches and cañons, at the mouths and far up the course of streams, below the level of lakes, and under the spill of mountain springs, each camp made its own law of the distribution of water, the modifications of prior right necessary to full development of the diggings, and such other matters as were necessary to the common use of this requisite agent in that industry. As a result there grew up the customary law of useful appropriation of water to the primary industry of the State. The Legislature of California, called to the duty of drawing around mining the circle of statute law, decreed that in all actions at law, concerning mines and miners' rights, the local regulations, the customary law of the mining district, should be the rule of decision for the guidance of the court. Here was the germ of the common law of California. Custom had laid down the law of location of mining claims, and custom had appropriated the water necessary to their operation, and the Legislature directed the courts to consider this customary law as their rule of decision. Now bear in mind that the Legislature had also, following the custom of senior States, enacted that the common law of England, where consistent with the constitution and laws of this State, should be the rule of decision in all the courts of this State, but the law of mining locations and water rights thereon, as declared by the same Legislature, was not consistent with the English common law of riparian rights, and therefore the customary law of the miners took precedence of the English common law, and was made the rule of decision for the courts.

This principle has subsisted, undisputed and undisturbed, from 1850 until the recent decision of the Supreme Court of the State. Perhaps it is not fair to say that it was undisputed and undisturbed, since its validity and authority were practically affirmed by that adverse possession which is a vital point in the law of tital by occupancy. Whenever the right of useful appropriation was assailed, it was maintained by our local courts, and it was affirmed and acquiesced in by the Federal and State Governments. In 1866, Congress for the first time legislated upon the subject. Sixteen years before, the State of California had made customary law the rule of decision for the State courts, and now her example was adopted as the Federal law, and made the rule of decision for the Federal courts, by this statute: "Whenever by priority of possession, rights to the use of water for mining, agricultural, manufacturing, or other purposes, have vested or accrued, and the same are recognized and acknowledged by the *local customs*, laws, and decisions of the courts, the possessors and owners of such vested rights shall be maintained and protected in the same." The first case that went to the bar of the Federal Supreme Court under that statute was *Atchison vs. Peterson*. Opinion by Mr. Justice Field, who was the author of the original law of California, making the customary law of mining districts the rule of decision of our State Courts. In this case he wrote this judgment of the Supreme Court: "By the *custom* which has obtained amongst miners in the Pacific States and Territories, where mining for the precious metals is had upon the public lands of the United States, the first appropriator of mines or of waters in the streams upon such lands for mining purposes, is held to have a better right than others to work the mines or use the waters. As respects the use of water for mining purposes, the doctrines of the English common law, declaratory of the rights of riparian owners, were, at an early day after the discovery of gold, found to be inapplicable, or applicable only in a very limited extent, to the necessities of miners and inadequate to their protection." The learned Justice then projects his argument into a demonstration that the appropriation of water for beneficial use had always "been heartily encouraged by the legislative policy of the State." Again, in the case of *Basey vs. Gallagher*, the Supreme Court of the United States, by Justice Field, referring to its prior decision in *Atchison vs. Peterson* said: "The views then expressed and the rulings made are equally applicable to the use of water on the public lands *for the purposes of irrigation*. No distinction is made in the Pacific States and Territories by the *custom* of miners or settlers, or by the courts, in the rights of the first appropriator, from the *use* made of water, if the use be a beneficial one. In the case of *Tartar vs. Spring Creek Water and Mining Company*, decided in 1855, the Supreme Court of California says: 'The current of decisions of this Court go to establish that the policy of this State, as derived from her legislation, is to permit settlers in all capacities to occupy the public lands, and by such occupation to acquire the right of undisturbed enjoyment against all the world but the true owner. In evidence of this, acts have been passed to protect the possession of agricultural lands acquired by mere occupancy; to license miners; to provide for the recovery of mining claims; recognizing canals and ditches which were known to divert the water of streams from their natural channels for mining purposes; and others of like character. This policy has been extended equally to all pursuits, and no partiality for one over the other has been evinced, except in the single case where the rights of the agriculturist have been made to yield to the miner where gold is discovered in his land.

Aside from this, the legislation and decisions have been uniform in awarding the right of peaceable enjoyment of the first occupant, either of the land or of anything incident to the land.' Ever since that decision it has been held generally throughout the Pacific States and Territories that the right to water by prior appropriation for *any beneficial* purpose is entitled to protection. Water is diverted to propel machinery in flour-mills and saw-mills, and to *irrigate land for cultivation*, as well to enable miners to work their mining claims, and in all such cases the right of the first appropriator exercised within reasonable limit is respected and enforced." Here, then, we have unmistakable recognition of the right of useful appropriation of water, asserted by and confirmed to the miners of this State, and as unmistakably we have the judicial *metastasis* of that right to the irrigator of agricultural lands, by an ascription as plain as legal reasoning can make it. The reports are crammed beyond the space at our disposal to digest, with decisions following the unvaried line of customary law as originated by the people of California in the necessities of those physical conditions peculiar to the State. This custom made the common law which the Legislature ordered the courts to make the rule of decision, instead of the conflicting common law of England, and under that custom appropriators' rights held adverse possession against riparian rights for thirty-six years of the legislative and judicial history of the State. It has become as much a part of popular rights and as entrenched in the public thought, and habit and custom as the right of trial by jury, the *habeas corpus*, or the elective franchise, and the recent decision, secured by a majority of one of the State Supreme Court, is as rude a blow in the face of public opinion as the court could have struck if it had swept trial by jury, *habeas corpus*, and the elective franchise into the abyss of a common ruin.

It is such a decision that has shaken the foundation of parties; that has engulfed all other public issues; that has painfully shadowed thousands of homes in the Sacramento and San Joaquin Valleys and Southern California; that has warned wholesale merchants in the city of the uncertainty of outstanding country credits and the instability of future trade; that has admonished railroads of a decreasing tonnage, their stockholders of diminishing dividends, their bondholders of defaulted interest, and that has notified city bankers that country loan accounts on real estate security are to be worth less than the paper spoiled in writing out the mortgage which did cover productive lands of marvelous fertility, transformed, by this judicial evil genius of the State, into a desert. No people ever permitted such a decision to stand, nor let the customary law, of thirty-six years' beneficial existence, perish by judicial assassination. No means known to the law and its processes, provided to reduce the judiciary to a condition of harmony with the physical necessities of the people, will be omitted during the pendency of this contest, and when it is over, the customary law of California will sit on the seat of authority, crowned and sceptered with supreme power, while the English common law of riparian rights will be returned to its own country as we send back pauper and criminal immigrants, who are not the material for good citizens.

CHAPTER XIX.

FRESNO COUNTY.

ONE needs to take a physic of figures to realize that California ranks next to Texas in size. Within its borders is room for three and one-half States the size of Iowa; England and Wales could be spread here and only cover one-fourth of the State, and whole States could be hidden inside some of its counties. Fresno County is an example, not only of area, elbow-room galore, but of a variety in soil and surface, mountain and plain, field and forest, vale and intervale and foot-hill, which equip it with all the physical characteristics and resources needed to make an independent political community. On the continent of Europe are independent nationalities that would not make one of its townships.

Fresno has an area of 5,600,000 acres; Rhode Island has 835,840 acres; so that Fresno could make six States as large as Rhode Island and leave a strip of 584,960 acres over for "cabbage." Delaware has 1,356,800 acres, so that Fresno County would hold four States like Delaware, with a margin of 172,800 acres left over for a nest-egg. Rhode Island and Delaware combined, have 2,192,640 acres, and both of them could be dumped into Fresno County twice and leave 1,214,720 acres over.

Switzerland is less than twice as large as Fresno, and twice Fresno would make another Denmark, with a respectable lap over. It is two-thirds the size of Holland and more than two-thirds that of Belgium. This lesson in size is needful, not merely to stimulate the territorial pride of Fresno's people, but to show them, and those who are to be of them, the great future that may be wrought out of such an area of such land as this county has. It is part of the San Joaquin Valley, and is traversed by the San Joaquin and King's Rivers, which rise in the Sierra Nevada Mountains, on the east side, and are fed to a perpetual flow by the rains of the wet season and the melting mountain snows of the dry. These streams furnish abundant water, and their delta especially offers unsurpassed facilities for irrigation; while the whole plain surface of the county is so situated that it may nearly all be reached by hydraulic works. On the west the Coast Range rises against the Pacific and shuts out the fogs from the sea. On the east, within the county line, is much of the most interesting of the high Sierra chain. Here are Mt. Whitney, Mt. Goddard and Mt. Lyell, amongst the loftiest peaks on the continent; Whitney rising to over 17,000 feet, and rearing his cloud-defying crest to the storms far above Mt. Washington and the noted peaks of the White Mountains and the Apalachian chain. Cradled between the Sierras and the coast range lies Fresno *Felix*, once tramped by bands of wild horses and cattle and bleating flocks, but rapidly changing under the magic of emigration and enterprise into a densely populated and rich region.

Out of its two rivers the clear mountain water, that sparkles with the sunshine it caught glinting through the pines and lofty cedars that flaunt their foliage far up the mountain, is taken in canal and ditch; and wherever it goes, grass and grain, grapes and olives, fruits and flowers, happy homes and wholesome people, are in its train. The sun of Fresno sought long the lucky sign. Around and round the zodiac it went in quest of the spell that should give to this great county a vision of the destiny it seemed to merit, and at last it stood still in the sign of Aquarius, the water-bearer. True, the Supreme Court of California has put its legal handspike into the spokes of the zodiac, to turn it back so that Fresno shall be again under the sign of Taurus, or Capricornus or Aries; but this will never be. Bull and ram have had their day on the dry plains, and Fresno will continue to conquer in the sign of the water-bearer.

Fresno copied the colony system of Southern California, and it has now, in productive operation, the Walters Colony, and the Scandinavian, Nevada, Fresno, Malaga, Central, Washington, New England, Belfast, Norris, Sierra Park, and Witham Colonies. It boasts also the celebrated Barton, Eisen, Eggers, Goodman, Forsyth, and Woodworth, Easterby, Mather and Fresno and Butler Vineyards, and the McNeil, Creek and other well-known commercial orchards. The capital, Fresno City, is about 200 miles by railroad from San Francisco, and has suddenly sprung into a well-built city of 4,000 people, and is growing with the rapid growth of the country around it.

Now, when we tell how this county has so suddenly supplemented its large size by great development, we tell the story of many other counties in California, as it is written already, or is to be writ in a speedy future. People resorted to the twenty-acre tracts of irrigable lands in Fresno, bought them on credit largely, put up a house, planted some alfalfa, kept a cow or two, built an adobe milk house, got some chickens, planted some vegetables and berries, and having begun by these means the process of self-support, devoted themselves to putting the rest of their twenty acres into vineyard and orchard. The vines, at three years from planting, began to yield, and thence on they yielded an income of from $100 to $300 per acre. That is all there is of it. The owner of twenty acres supports his family and puts in bank, every year. $1,500 to $2,000. In the agricultural States of the East he would not do that on a half-section of land. Here he is in a winterless country, with two seasons, the wet and dry. The first is spring, the second is summer. It is the ideal raisin climate. The air is dry, the sunshine converts the sweet juices of the grape into that spicy jelly which is the test of this king of dried fruits, and though the days are hot, at night the cool winds come down from the snow-capped mountains, and the farmer is called from labor to the refreshment of sleep in blankets.

We treat Fresno at length, as a typical valley county, illustrating the results of industry and irrigation. For field crops it produces wheat, corn, Egyptian corn, potatoes, sweet potatoes, peanuts, sorghum cane, and its orchards are made up of pears, peaches, apricots, nectarines, prunes, plums, oranges, lemons, and the olive. The wines of the county have already established a high character. Now what more do you want? Immortality? it must be in the impression your stout hand writes upon this enduring page which Nature has opened to record the exploits of thrifty men.

CHAPTER XX.

TULARE COUNTY.

THIS is the fourth county in California in size. It lies midway down the San Joaquin Valley, in the middle of that great pocket which is turned upside down to empty its contents into San Francisco. South of it, in the bottom of the pocket, is Kern County, and north of it is Fresno, which it resembles in its mountain boundaries and topography. Its hydrography, however, is peculiar to itself. The waters of King's, Kaweah, Tule, and Kern Rivers flow in Tulare. These rivers and streams head in the Sierras, and during the dry season are fed to flood height by the melting snows. There seems to be some providential interposition in this fact, some law older than the common law of England, for the flood of mountain water fills the streams at the seasons agriculture and horticulture most need irrigation. As the riparian rule is finally relaxed, and we have law fitted to our natural conditions, the flood waters of the rainy season and the melting season will all be impounded behind dams at their mountain source at proper intervals along all these streams, and then its volume will be found ample for perfect irrigation of every part of this noble principality upon which a home can be founded. There is nothing stronger than man's attachment to old ideas. The farmer who carried his grist in one end of the sack, balanced by a stone in the other, is not a mere figure of speech. The history of various useful inventions proves this farmer to be no myth. Witness the model of the first reaping machine, which was a great disk with scythes set in its rim, to which it was intended should be given the motion which the hand cradler gave his scythe through its snath. The inventor could not give up the idea of reproducing the manual motion, but he was followed by one who thought out the sickle bar, with its toothed guards and reciprocal motion, and the problem was solved. So the model of the first threshing machine is an affair run by oxen in a tread-mill and arranged to fling flails, in imitation of the manual motion given that primitive implement on the threshing-floor. When this clumsy machine was set going, by some twist in a belt or squint in a cog-wheel the flails turned upon the oxen and beat their horns off. It was a failure, but soon a man who could put old ideas behind him, invented the toothed cylinder with its complementary concave, and the flail went to join the scythe in disuse. So it was the old idea that we brought from England, that the soil must be moistened by rain, and that rivers are for navigation, fish, water-power, and drainage purposes. To use the water for irrigation has even been held by many good people to be sacrilegious, and in one New England church the digging of a canal was opposed, because it was said that where God intended water to be there he had put rivers and springs, and only a man of sin would defy Divinity by moving to amend in any way the arrangements of Providence. One brother, who had an interest in the canal, carried it by quoting scripture, "And Jacob digged a well."

So the pioneer irrigators of California had to fight the tradition that the plains where Providence withheld its watering pot had upon them the primal curse of barrenness. It was the same as if man's dominion had ceased at the seashore, and he had never built a ship for discovery of what lay below the ocean horizon. The irrigator remembered that rain fell on his father's fields, and when it failed then crops were gnawed by the drought, and had this memory dominated him he would never have digged a canal nor brought land and water together, subject to his will.

Tulare has 4,100,000 acres of land. The quality on the east side, near the Sierra foot-hills, is gravelly and adapted by irrigation to fruit culture. Within a few miles, upon the plain, it changes to a dark, sandy loam, a "quick" soil which gains in richness continually as the river bottom and deltas are approached. Scattered over the county are alkali lands, once thought to be worthless, but proved by cultivation to be strong and excellent soils. Tulare used to be a "cow county," given up to bands of live stock, and supposed to be worthless for agriculture. The late Col. John C. Hays, the Texan ranger and one of the most charming characters in all our frontier history, was one of the earliest believers in the capacity of this county, and he did much to encourage the prosperity which is now coming to its hardy people. Its productions are the same range as those of Fresno, and it has the same capacity for supporting a dense population on small holdings under the colony system; but it will probably remain the seat of a vast grain production much longer than its neighbors, Kern and Fresno; and its live stock interests, supported by alfalfa, will always be among the permanencies of the county.

Now, how can we impress a home-seeker with the opportunities which await him in Tulare? The prevalent Eastern idea is that there is no new country in California. This State began to be talked of during the war with Mexico, while Iowa was a Territory, and before Minnesota, Kansas, and Nebraska were organized as Territories. To the Eastern fancy, unenlightened by exploration, the process of occupying our lands has been concurrent with the occupation of those of Iowa, Minnesota, Kansas, and Nebraska, and if so, they are all taken, and no virgin soil offers fresh opportunity to the new settler. But these Eastern people forget that so far as occupying the soil for tillage is concerned, California is now, in 1886, about where Iowa and all that region was in 1856 Why, it is only within five years that the large land-holders of Fresno began to divide their estates and part with their principalities to make a chance for colonial settlement; and these lands, which in their raw state five years ago were to be had for $2.50 per acre, are to-day paying their tillers interest on a valuation of $200 per acre and upwards. Now this process is just beginning in Tulare, with the same climate, a soil of equal character, and, as we have said, a hydrography that is ample for perfect irrigation; and men who have witnessed the mighty things wrought in Fresno are rapidly seeking locations in Tulare, for twenty-acre tracts can be had there now for $250, of which $50 is paid down, and the balance in monthly installments of $10. A laboring man ought to soon save up $50, and he ought to earn a surplus of $10 per month for twenty months to secure a home in Tulare. After that, with the land under his feet for a basis of credit, he ought to be able to plant some alfalfa, get a hog, a cow, some chickens, and a shanty to begin on, and he ought in a few years to be independent. They have done it in Fresno on an original cost twice as great, and many a twenty-acre tract there begun just this way, is now worth anywhere from $4,000 to $8,000. The pioneer in the upper Mississippi Valley, who faced hail and lightning and cyclone and rain all summer, and an arctic temperature all winter, met more obstacles and endured more hardships in any one month of his noviitate than a new settler will find in five whole years in Tulare, and by that time he should be in the way, off the produce of 20 acres, of putting $1,000 a year in bank, over and above all the expenses of his family. If a settler on a quarter-section in the prairie region east of the Rocky Mountains, is able to do this much after twenty years of hard toil, he is in luck. That it is done on four years' developments in California we can prove by cases so numerous that the citation would fill this book from lid to lid; but this is not a real estate circular, and we give only outline facts that can be certificated over and over again.

By this time the reader wants to know how the name Tulare is pronounced, for " he wants to go there too,"

> To try Tulare,
> Bright and airy,
> Where the fairy
>   Might herself find a home.
>
> Where no frost nor snow,
> Nor icicles grow,
> And gentle winds blow,
>   For cyclones never come.

Visalia is the county seat, but Tulare City and Hanford are thriving towns, and reached through the Southern Pacific Railroad system.

CHAPTER XXI.

KERN COUNTY.

THE San Joaquin Valley is the valley of the Nile in miniature. Its lower portion, approaching the Bay of San Francisco, was originally susceptible of grain production, in most seasons without irrigation; but as you ascend the valley it grows more arid until in Kern County, at the extreme tip of the pocket, you have reached the Soudan of this little Egypt. It was, for the most part, originally desert, much like that hopeless stretch between Wady Halfa and Uganda, down which Chinese Gordon rode camel-back to Khartoum, and out of which, disembodied, he took a mount behind Death, on the pale horse. True, Kern has valleys, and they were early occupied, but this occupation gave the county but little importance. Between Kern River Slough and the old channel of that river is a great delta, that had been an impediment to the development of the county. It was arid, desert-like, hard to traverse, a little Sahara, dry, thirsty, and as unprofitable a part of the footstool as could be found. This unpromising delta was occupied by J. B. Haggin & W. B. Carr, the first men to extensively and comprehensively apply irrigation to redemption of lands so extremely lost in the original sin of unfruitfulness as to require more faith than would move mountains to back up an effort for their redemption. Any man who wishes to indulge in the most fascinating study of

HEAD OF KERN RIVER.

this question that can be made in the world, should go to the results in Kern County, and the means by which they were accomplished. Poetry has exhausted its metaphors on the sculptor whose voluptuous fancy sees a Venus in the rough block of marble, and whose hand delivers her from its imprisonment, but a far finer subject is offered in the men who saw in this gleaming desert, farms, vineyards, orchards, and homes, shade and shelter, flowers and fruits, as the concrete of those visions with which the glimmering mirage had lured the traveler to disappointment. The canals and ditches of this system have carried water far out upon the plains, and wherever they go green fields and prosperity have followed. We have spoken of the San Joaquin Valley as a pocket. It is, and so is the Sacramento Valley. They lie in each groin of the State and are open toward San Francisco. Kern is at the bottom of one pocket, and as the coin is always found at the bottom of the purse, here it is. The county is a most interesting region. Dairies and stock farms lie green and cozy under its ditches. Its productions run from cotton through all the cereals, fruits and grasses, to root crops. Its irrigation facilities already in operation supply 677,000 acres. Some of its canals are 150 feet wide, and so this Soudan of the San Joaquin Valley has been conquested, and the peaceful conquerors have occupied it, and the commerce of San Francisco rattles the consequent coin in the bottom of this right-hand pocket of the State.

Mountains are on three sides of Kern. The Sierra Nevadas meet the Coast Range at an obtuse angle, and their spurs push far out upon the plain, great buttresses and pilasters holding up the granite wall, which rises from 2,000 to 5,000 feet on all sides but the north. Kern is the *ultima Thule* of the San Joaquin. Passing its mountain sentinels, you go into another region, with its own mountain and river system and its physical peculiarities; but Nature dictated that from Kern the flow of wealth which the industry of man shall generate, should run to the Golden Gate.

The area of the county is 5,137,920 acres. Its capital, the Khartoum of this regenerated Soudan, is Bakersfield. Within its great borders the colony system is fast making its way. Mr. B. Marks, of San Francisco, has brought to bear, upon the problem of colony location, his great experience acquired in other parts of the San Joaquin Valley;

and from this time the resources of this distant contributor to the common wealth will be rapidly developed, as its merits are made known to those who seek such a happy combination of climate and soil, and love the wedding garlands which festoon the marriage bed of land and water.

FRESNO, TULARE AND KERN.

We have dealt with the three great counties, Fresno, Tulare and Kern, in a group because their characteristics are harmonious, and the people who are and who are to be in possession of their soil and in enjoyment of their climate, will find themselves so affected by common interests as to keep step with each other in nearly all matters material to their welfare.

Together this group of irrigable counties presents an aggregate area of 14,837,920 acres. It has within its limits the climate of Italy, the scenery of Switzerland and it throws the products of Southern Europe, while its homes are bowered in the surroundings of semi-tropical Asia. Measured in square miles, the area of this trinity of counties is 23,184. Belgium is only half as large, and has a population of 5,800,000. Denmark is only two-thirds as large, and has 2,038,000 people. Greece, with the islands and Thessaly, has the same area, with 2,120,000. Holland is only half as large, and has 4,280,000. Switzerland is only two-thirds as large, and has 2,930,000. These counties are half as large as England, with 27,500,000 population, and they lack but little of being as large as Scotland, with 3,900,000 people, or Ireland, with 4,950,000. So it will be seen that these counties are not crowded, for they do not yet contain 75,000 people; but as almost their entire irrigable area can comfortably support a family on each twenty acres, it will be seen that this is destined to be the most densely populated, as it is naturally the most fruitful, part of the continent.

Why should not a proper circulation of the tidings of promise rapidly fill such a region with a prosperous people?

Ten years ago there had been no appreciable beginning made in Fresno, Tulare and Kern. Now they are dotted with settlements which demonstrate their capacities. Ten years hence they will be known for their progress and thrift. Threading them runs the Southern Pacific Railroad, and it will soon bristle with side lines and feeders.

Why should people snub such a country to go to Australia, toward which so many colonists longingly look? The London *Standard* in a late issue says that South Australia has just raised another large loan, which makes her the peer of Queensland and New Zealand in the burden of public debt, which now averages in the three colonies from $250 to $300 per head of population! On top of this comes the news from Sydney that the deficiency in the exchequer for the past year is $8,500,000, which is to be met by the imposition of a larger land taxation, an income tax and an additional tariff of 5 per cent *ad valorem* upon imports!

California is the rival of British Australasia as a field for colonization. Here is no crushing debt, no increasing land and income tax, no progressive deficiency in public finances to be made good by wringing the withers of labor and robbing production of its profits. Our public debts are decreasing, our taxes growing less per capita. The contrast blows its own trumpet.

CHAPTER XXII.

SOUTHERN CALIFORNIA.

LOS ANGELES, SAN BERNARDINO, AND SAN DIEGO.

SOUTHERN California! Books and poems have been written and lies have been told about it. In the Eastern fancy it is of dreamy outline, and all manner of tales go touching its crops, its climate, and its people.

We have grouped the three great counties of the San Joaquin Valley together, for Kern, Tulare, and Fresno are joined in the same destiny, thrive or shrink together; and the wealth that is in them waiting for thrift and enterprise to develop it, goes directly to San Francisco. In one respect these counties and all those of the San Joaquin and others in all parts of the State have common cause with Southern California, with that part of the State below the pocket of which Kern is the bottom. Irrigation is the tie between the three great counties of the south, Los Angeles, San Bernardino, and San Diego; and the rest of California, albeit their wealth is not poured primarily into the lap of San Francisco, but is going into the building up of a new Baltimore and Philadelphia. We would call Los Angeles a Pacific Chicago, except that its name might be at odds with the designation, if we take into consideration the reputation for other things than enterprise which distant people ascribe to the City of the Lake. But it must be remembered that a city resembles a man, in this, that success is supposed, by the ignorant, to imply a knowledge of magic and the black arts, when, in fact, success comes of knowing and minding your own business.

Los Angeles County has been a marvel of progress and enterprise. In some of her older wineries are wines from vines that were planted, cultivated, and had their grapes picked and pressed by Indian labor, employed by the early settlers. From such a beginning she now possesses the greatest winery in the world, which, under the management of its executive head, Mr. J. De Barth Shorb, turns out annually the largest number of gallons made in one establishment under one head. In the Nadeau Vineyard she now has also the largest vineyard in one body, under one ownership, in the world. The territory which now makes this county was settled in 1771, at San Gabriel, by the mission fathers. At the conquest, the pueblo of Los Angeles was the Mexican capital, and there the last Mexican Governor, Don Pio Pico, still lives, verging upon a hundred years old. He is the Petrus Stuyvesant of his people, for old Peter, when compelled to give up New Amsterdam to the English, took his revenge by chopping down the English cherry trees in front of his house and refusing to learn the English language. Don Pio is a monument to the pride and steadfastness of the Mexican character, and is an interesting and suggestive figure of the past, in the midst of the surging life and vital enterprise of the present. Los Angeles County is within a third as large as the State of Massachusetts, having 3,600,000 acres. Its southern boundary is San Diego County and the Pacific Ocean, on the north is Kern; on the east, San Bernardino, on the west, the Pacific again and Ventura County. The bounds fixed for it by nature, which determine its climate and enrich it with great capacities, are the mountains and the ocean. You leave the bottom of the San Joaquin pocket, cross the Tehachepi Mountains, which connect the Sierra Nevadas and Coast Range, and after passing such marvelous triumphs of civil engineering as the Loop, where the railroad crosses itself to climb the difficulties of grade, you slip down into the verdant, blooming, and teeming meadows of a country that has upon one side the Sierra Madre Mountains, and on the other the sea. This is Los Angeles County. Here were laid the foundations of the New California. Here the problems of irrigation were worked out for the benefit of the whole State, and here was developed an orange belt which is to supply 60,000,000 of people with citrus fruits in the interval between the Florida and Italy crops. Here the grasses flourish, from those which herds graze or pasture, to wheat, rye, barley, and the noblest grass of all, Indian corn. Beans and potatoes, the sugar beet and all root crops, clover, hemp, flax, melons, pumpkins, and berries reach a perfection possible only to such a soil and sun, joined to useful irrigation. But a few years ago this was a cow county, where the holders of old Spanish grants lived the ideal ranche life, with their haciendas, the home of all their people and dependents, as was the castle of a Scottish chief the home of his clan two hundred years ago. Here the major domo saw to it that none who belonged below the salt should sit above it at table, and in the dreams which whiled away each *siesta* the then lords of the soil saw no vision of what was to be. Into this land of the lotus eaters came the enterprise of the immigrant and capitalist. The dry plains and *mesas* which had grudged a lean pasture to sheep and cattle were transformed, as by magic, into orange groves and vineyards. The waters of the Rivers San Gabriel, Santa Ana, and Los Angeles were harnessed to the plow, and the attractions of climate were soon supplemented by the verdure and fruit of our enlivened landscape; and the result is a delicious series of rural settlements, than which nothing can be more attractive The shores of the Mediterranean have nothing to offer that can surpass the blandishments here, except historical associations, and what does the dreamer in a Los Angeles hammock care if he cannot look out upon the scene of the Sabine rape; and does the vineyardist of San Gabriel, or the orange farmer of Pasadena, enjoy less the profits that come out of the soil because it was not fattened by the dust and bones of noble Romans? Around the Mediterranean such scenery and its historical associations are partners with age and industrial decay. Commerce left those shores and sailed out between the Pillars of Hercules long ago. Here in Los Angeles is the thriving, bustling capital of that name, a marvel of trade and activity, with its markets dealing in the fruits, wines and oils, nuts and raisins, figs and pomegranates, which we associate with our ideas of the trade of Palermo and Nice. Around it, as mountain snuggeries, or seaside resorts, or jewels set in plain or foot-hill, are San Fernando, Pasadena, Sierra Madre Villa, San Gabriel, El Monte, Duarte, Azusa, San Pedro, Santa Monica, Santa Ana, Spadra, Downey, Cerritos, and other suburbs and rural places and colonies, each with its own attractions, as in a family of sisters each may have graces of her own that detract nothing from the rest, but give her zest in the eyes and arms of her lover.

The first impress made by civilization here, as we have said, was at San Gabriel Mission, from which all that is now Los Angeles County was ruled. Referring to the ancient mission census, we find that under the padres the county had 105,000 head of cattle, 20,000 horses and mules, 40,000 sheep and goats, and produced 20,000 bushels of grain in a year. It was truly a cow county, but how surprised the pious fathers would be at the change that has come over their grazing grounds. By the assessment of 1885 it had 27,070 head of cattle, 15,568 horses and mules, but it had ostriches on the plume-raising farm at Anaheim; it produced wine to the value of $5,400,000; it exported 139,000 boxes of superb raisins; it filled about half of the 1,000 car loads of Southern California oranges that went east to be sucked by our countrymen; it exported to Europe 20,042,397 centals of wheat; and its population of about 100,000 is increasing so rapidly, through the channels provided by nature and immigration, that it is hazardous to venture figures.

San Bernardino has many characteristics in common with Los Angeles, and San Diego, with its imperial area of fifteen millions of acres it is capable of the happiest transfiguration by irrigation and enterprise. These three make up that Southern California—that land of the orange and the vine—which has a magnetic fame that reaches around the world. In London we have heard a noble lady say, "I *must* see Southern California again," and in Paris a *blasé* man of the world cried out in its praise, "Ah! how like France!"

This part of California is what all the State is to be. Be sure of it, the results that have been conjured by enterprise in this fairy land are, like faith, the substance of things hoped for all over California and, with harmony between natural necessity and statute law, the hope will prove to have been not in vain.

CHAPTER XXIII.

MERCED COUNTY.

CONSIDERING the richly-lined pocket of the San Joaquin Valley, the counties that lie above Fresno have peculiarities that are notable. The first is Merced, with an area of 1,155,336 acres, of which three-fourths is susceptible of profitable cultivation. The capacity of Merced has been shown in those seasons of abundant rain-fall which have given the soil all that it needs of water to show its fertility. In seasons of low rain-fall, production recedes, and the margin between shows the value of permanent irrigation applied to the acreage already under tillage. Take this margin of difference upon the produce of one acre and multiply it by the irrigable acreage of the county, and you get the annual money value of irrigation to this one county. In a season of full rain-fall the average yield of wheat is 30 bushels to the acre. In the years of low rain-fall it is nothing. Suppose that the total tillable area of the county were under the plow, what is the loss represented by this lack of irrigation, if the land were all in wheat? The tillable area being 866,451 acres, at 30 bushels per acre the loss in bushels is 25,993,530, and at only 60 cents a bushel, the loss to the county in money, in one year, for lack of irrigation, would be $15,596,118. But wheat is far from being the most valuable crop that Merced would produce if its climate and soil were permitted to do their best by adequate use of the abundant waters which might be taken from her streams or impounded in the mountain cañons for the benefit of her fields in the dry season. The Supreme Court of California, by its recent decision, says to Merced County: "You shall not have the settlers to plow your glebe and produce this wealth. You shall go on losing fifteen millions a year that might be earned by the owners of homes and farms and vineyards and orchards on your soil, because English law, which is our rule of decision, is not favorable to your prosperity." It will be seen that the people of Merced will show scant mercy to such a law and to such a court, when they can in any way influence a reversal of the one and a change in the other.

The San Joaquin and Merced Rivers flow through the county. The latter has its rise far up in the snow above the Yosemite, and it winds through that valley on its way down to the plains below. The Merced is rich in water-powers as it comes down through the foot-hills on the east, and as the county develops, the water, which is finally to irrigate the prosperous farms, will first have turned the wheels of many a mill and factory, and out of this double duty will come a duplicate profit to happy Merced. The surface of the county is picturesque. There are many groves of live oaks, and the vines and fruits planted there thrive in all parts. The county seat is Merced City, on the Southern Pacific Railroad, 151 miles south of San Francisco; and its fine hotels and dry and bracing climate have already made it a resort favored by many who seek health or pleasure in the vacations from business. The same mountain ranges that line each side of the whole valley make the boundaries of Merced.

Capital has just been attracted by the county's latent resources. Mr. Chas. Crocker and Mr. C. H. Huffman have tapped the Merced River at Snelling, at a level so high that by cutting a costly tunnel the water is carried out above the plains for a distance of 35 miles in a canal 80 feet wide at the bottom and 100 feet wide at the top. It will irrigate perhaps half of the irrigable area of the county and some land in Fresno. It is an enterprise of wonderful magnitude, and its possibilities were discussed for years by the pioneers of Merced, who despaired of its accomplishment. Now it is a fixed fact, and the owners of the great tracts of land to which it will bring water are preparing to subdivide, and to invite the colony system, which has done so much in Fresno, Tulare, Kern, and Southern California. Besides this grand canal and the streams mentioned, there are, in the county, the Chowchilla, Deadman's, Mariposa, and Bear Creeks, and their tributaries, all considerable streams and of value in considering the comprehensive irrigation of such a vast and valuable body of land. There are other irrigating systems on both sides of the San Joaquin, and they are the means of showing, in flattering and favorable light, the capacity of the county.

Let it not be imagined, however, that Merced is all in the future. The institutions of civilization and society have long been founded there. Forty district schools show an attendance of 89 per cent of the little folks of school age in the county. The assessed valuation of property in Merced, real and personal, is $12,322,224, which for a population of about 6,000 is an evidence of wealth-producing capacity that tells its own story to the intending settler; and when it is remembered that only recently all this land was ranged by sheep, and that still later wheat and wool were the sole factors in its commerce, the diversity of crops produced by irrigation may well surprise the old-timer. Soon the wines and raisins and fruits of Merced will take a distinct standing in the market, and the county will be covered with thrifty colonies.

STANISLAUS COUNTY.

Next below Merced lies Stanislaus, which runs from the Sierra Nevada foot-hills on the east to the summit of the Coast Range on the west, completely spanning the valley. Its neighbors on the east are Tuolumne and Calaveras; on the west it is bounded by Santa Clara. Its area is 924,800 acres. The San Joaquin flows through it, and receives as confluents the Stanislaus and Tuolumne Rivers. Bret Harte has immortalized the Stanislaus as the scene of an adventure of "Brown of Calaveras."

Stanislaus County has been a great wheat producer. The crops raised on the east side of the San Joaquin have been marvelous in years of average rain-fall, and they seldom fail entirely. On the west side of the river, however, owing to the influence of the Coast Range, which sends the rain-laden clouds sailing too high for precipitation, the crops are far between; but so rich is the black soil that land owners there say that a crop one year in five pays them. This being true, fancy the production that will be fostered by irrigation. Already the San Joaquin Canal serves about 20,000 acres of this land, where the soil is a black loam, from 10 to 100 feet deep. This canal will finally be extended along the west side, carrying water to about 90,000 acres of these lands which are shunned by the rains, but have all the other conditions of fabulous production.

The farmers of this county have observed the good results of diversifying rural industry, though it is hard to change from the habit of wheat farming, and resort to irrigation and seems a change which dismays the conservative. Tradition will not hold back the new-comers who have seen the mighty things that have been done in Southern California and the three lower counties of the San Joaquin Valley. Six times since 1850 the whole county has been parched by stubborn droughts. The loss of stock was enormous, and the crops totally failed. These experiences admonish to diligence in introducing more extensive irrigation from the waters which abound in the streams, and in this necessity the people of Stanislaus have a community of interest with their neighbors above.

The county can, under easily controlled circumstances, produce as great a variety of fruits, grains, roots, and nuts as any part of the world, and its red lands in the eastern foot-hills, now producing two crops of wheat in three years, have the finest adaptability to the grape, and will produce sound and standard wines.

The county seat is Modesto, a thriving town already, and destined to greater growth, as manifest destiny has its sway in the country that will pay tribute within its gates.

CHAPTER XXIV.

THE SAN JOAQUIN VALLEY AND ITS MOUNTAIN RIM.

WE began our examination of the San Joaquin Valley in the colonies of Fresno, and thence have gone up the valley to Kern County. We now reach the limit in the other direction. We are at the mouth of the pocket. San Joaquin County is at the beginning of the valley. Its plains stretch to the valley of the Sacramento. This county has an area of 928,000 acres, of which only 51,813 are waste land in the rivers or too broken for tillage. The San Joaquin River divides into three channels in this county, and so makes some of the largest islands in the State. The Mokelumne and Stanislaus Rivers are confluent with the San Joaquin in this county, and the three streams supply abundant water for irrigation. This is one of the most important counties in the State, producing fruits, grains, and root crops, and devoted greatly to fine stock breeding and the dairy. Its products reach a final market by water or rail, and its farmers have prospered and list high in financial institutions. Stockton is the capital town, a great grain market, and the second city in the State in its manufactures. Being in the center of so great a grain country, Stockton is naturally the seat of a great trade in field and harvest machinery. Here is made the Shippee harvest machine, which cuts, threshes, sacks, and delivers the grain as it goes through the field. There are also foundries, tanneries, wagon and carriage factories, and the germs of a general manufacture fitted to the needs of a large population. Stockton is at the mouth of the valley; it is the Cairo, as Bakersfield is the Khartoum, of this New Egypt and Soudan. When the San Joaquin Valley shall feel throughout the impulse of enterprise, and be brought into productiveness by the skillful use of the waters that now waste in its streams, or evaporate in its lakes and ponds, or breed fever in its marshes, Stockton will be a great city. Its manufactures will multiply and the wines, brandies, raisins, figs, nuts, olives, oil, and grain of all the valley will contribute to its trade. Trains will load for Stockton far up the valley, and as the draft or bill of exchange follows the bill of lading, Stockton banks and moneyed institutions will keep pace with the addition of every acre to the producing surface of the valley.

The land-holders of San Joaquin County are fired by the example of Fresno, and, conquering their pride in vast possessions, are dividing their estates and inviting the settlement of colonies. They are also encouraging comprehensive irrigation, which will greatly increase the productive capacity of their lands. Considering the position of this county and of its chief town, its people have a present interest in putting irrigation amongst the permanent policies of the State, that should rouse them to powerful exertions in that behalf. While San Francisco is the New York of the New Empire, towns like Los Angeles, San Diego, Sacramento, and Stockton are, or are to be, its Bostons, Philadelphias, and Chicagos. As the rivers flow from far Kern with ever-increasing volume until they carry the commerce of Stockton out to the bay, so the commercial results of productive enterprise, fostered by the beneficial appropriation of water, will accumulate as they come down the valley, each colony and each city taking and contributing a share, until, when Stockton is reached, the produce of the valley will turn the wheels and gild the spires of a great city.

The four riparian judges of the Supreme Court propose to hinder, perhaps prohibit entirely, all this rural and commercial development. They fling the pall of English law, not only over all this valley, but their decision profoundly concerns the outlying counties; for on each side of the San Joaquin Valley are foot-hill and mountain counties whose future is to be greatly affected. It concerns them whether they are to be upon the borders of a desert again, or are to look down on the fairest and most fruitful valley in the world, and to share its prosperity. Next to the ocean or the bay are Ventura, Santa Barbara, San Luis Obispo, Monterey, San Benito, Santa Cruz, Santa Clara, and Contra Costa, each with some internal need of irrigation, each to be benefited in its own production by the use of water, and each with a great stake in the prosperous future that is now condemned to float out to sea, by these judges. On the east side of the valley are Inyo, Mono, Mariposa, Tuolumne, Calaveras, Amador, and El Dorado, mountain counties, the reservoirs of the State, within whose limits the rivers take their rise, and whose cañons and gulches offer the cheapest facilities for impounding storm waters to be found in the world.

These counties are rich in resources that the useful appropriation of their own waters will develop. The volcanic soils of their foot-hills will rival the steep hills of Bingen in their wine; and on these first steps of that mountain stairway by which we climb out of California, for 700 miles there will one day be an unbroken line of vineyards, orchards, and gardens, unequaled in the world for the variety and character of their productions. Think of the San Joaquin Valley set in such a frame! with mills and manufactures founded on the fine water-powers furnished in all these mountain counties, and with a reciprocal prosperity which makes the desert to blossom as a rose, and sets the mighty feet of the mountains in the midst of vines and orchards! The people who foresee all this, who have seen other States grow from nothing to millions by making wise use of their natural advantages, will have but little patience with four riparian judges who decide and declare that California shall not make any use at all of her most obvious, plentiful, and valuable natural capacities. It is as if this court should order that every merchant in this State shall do business with one eye closed and the other darkened by a colored eyeglass; that every blacksmith, painter, carpenter, mason, tailor, harness maker, and handicraftsman shall work with his left hand only, while his right is strapped helpless to his side; and that every farmer shall hold the plow with one hand and follow it on one foot. Such a decision, as the reader will see, would be equivalent to depriving the people who toil and produce, of one-half their natural capacity. This riparian decision does just this for the State, and the public welfare and respect for law demand the success of the mighty movement now in progress for a legislative reorganization of the court that shall bring new and unpledged judges fresh from the people and reflective of the public will and the popular welfare.

CHAPTER XXV.

AROUND SACRAMENTO VALLEY—THE MOUNTAIN COUNTIES OF THE NORTH.

WE have considered the line of irrigable counties from Mt. Diablo to the Mexican line, and the mountain counties which border them, and have an interwoven destiny. We will now look at some of the counties whose rural industries do not absolutely and primarily depend upon irrigation, though all their commercial interests, manufacturing activities, and financial institutions are touched at all points by the development of the State, which is amongst the certainties to follow the beneficial appropriation of water. Marin County is the mountain gateway to a region rich in interest to the capitalist and settler. Marin itself is all mountain, with the high valleys found in such a region. Lofty Tamalpais rears his frowning front in the center of this county, and is at once the scenic attraction and the water reservoir for the slope that faces the bay and that which looks out upon the sea. It is a dairy county, and the Italian and Swiss people have found congenial opportunity for that industry which they learned in their native Alps and Appenines. The county seat, San Rafael, is the ideal suburban home, and here live hundreds of business men from San Francisco, attracted by the charming scenery and clement climate. Here the fig and grape reach perfection, and the orange ripens its fruit and scents the air with its bloom.

Going through Marin by the Donahue railway, you enter Sonoma County. The Russians made a lodgment up here about the time they were capturing Alaska; but they left no trace behind, except in the name of Russian River, which brawls down this charming valley. Sonoma has about one million acres, with a frontage of sixty miles on the Pacific Ocean, and eighteen miles on San Francisco Bay. Here was the seat of Mexico's military power before the conquest, and in old Sonoma City still lives, in an honored old age, the Commandant-General Vallejo, who has been foremost amongst the progressive citizens of that State once ruled by his sword.

The Russian River Valley is rich in every resource of agriculture and horticulture. It is a land of corn and wine, and its charming foot-hills are flaunting the silvery green of great orchards of olive trees. In this county are medicinal springs that have done much to make people acquainted with its beauties and its bounties. Its apple orchards are the finest in California, and its grapes and wines are long celebrated. The county seat, Santa Rosa, is gemmed with all the charms possible to such a glorious climate; and Healdsburg in Sotoyome Valley, the vale of flowers, is surrounded by rich farms. The railroad terminates at Cloverdale, a lovely village worthy a rhapsody of its own, and a branch goes to Guerneville, in the heart of the Sonoma redwood region. Sonoma is always visited with delight and left with regret. The railroad which pierces its great valley has been a prime factor in its development, and the builder of that highway of its commerce, the late Peter Donahue, will hold a first place always in the esteem of its citizens.

Sonoma's neighbor on the north is Mendocino, made up entirely of two ranges of the coast system of mountains. Its wild grasses have made a favorite sheep and cattle range, and in its cultivated valleys the production of hops has been a profitable industry. Many a bold Briton has slaked his thirst with ale and stout that got its tonic bitter from the hops of

Mendocino. The Russian River rises in this county, and its valley shows some rich land, while the universal growth of wild oats and clover shows that when the lumbermen have stripped off the redwood, pine, fir, oak, and madrona, here will be the home, and breeding ground of fine stock and the seat of a fine dairy industry. The value of the forests may be seen in the fact that the redwood timber covers 745 square miles and redwood, take it all around, is the noblest and best tree that grows.

Next north is Humboldt, still in the same tier and on the coast. It is believed that when Sir Francis Drake sailed this way, he anchored in Humboldt Bay, out of which now goes and comes a commerce which would have been a richer prize than the silver-laden Spanish galleons of which that grim sea-king was so fond.

Humboldt County has an area of 2,400,000 acres, of which 640,000 are covered with the majestic redwood forest. The whole surface is mountainous and rugged, watered by Trinity, Mad, Eel, and Mattole Rivers and their confluents The Coast Range rises here to lofty peaks and throws its spurs out toward the sea. To see the standing timber is alone worth a pilgrimage across the continent; for it is one of the wonders of the coast, and one of the most reliable sources of wealth and profit that the New Empire can boast. Enterprising lumbermen have here overcome the difficulties of logging offered by the climate, or, rather, they have converted those difficulties into facilities. In the pine forests of Michigan and Wisconsin the deep snows permit the sledding of logs to the stream which is to float them to the river, where they are to be compacted into rafts. The same snows melted supply the water that is to carry the log on its journey to the saw. In Humboldt there are no snows to sled on, so enterprising lumbermen have built a logging railway, which runs into the timber belt. As the forest is on the mountain slopes, the great logs, when cut, are easily rolled to the specially made cars of this railway, and on them hauled to the mill. The Humboldt mills have a direct trade in lumber with the Pacific islands, and look forward to the penetration of the county by a railroad, that they may have direct shipment to the East. Redwood is rapidly taking the place of walnut and mahogany and rosewood in fine finishing and furniture, and its rich tints make it unnecessary for it to sail under any colors but its own. It furnishes the most beautiful veneers from the whorls of its stumps and roots; and as the roots do not decay, the loggers are leaving behind them in the ground a source of wealth that will soon be sought after by gangs of stump pullers. Although mountainous, the lands of Humboldt are surpassingly rich, and where the loggers have made clearings, it has an agricultural future. Around the county seat, Eureka, there is level land, and in the river valleys the land is so black and rich that it is called "niggerhead." Here the grasses, tame and wild, flourish, and it will be one day the "blue grass" region of California.

We are now in the most unbrokenly mountainous country on the continent. The counties of Del Norte, Trinity, and Siskiyou present noble mountainous boundaries, rich in mines, with gentle valleys that are prized by their fortunate occupants. Here is a hardy and honest people, as devoted to their mountain homes as the Swiss and the Tyrolean. Modoc County has only recently been wrested from the pest of Indian occupancy, and, like its neighbor Lassen, on the south, waits for railroads to fully develop resources that are important to the future of the State. Lake County, lying east of Sonoma and Mendocino, is but recently settled, and is still isolated by lack of railroads.

It is the opinion of Judge S. C. Hastings that this county offers excellent inducements to people who cannot reconcile themselves to artificial irrigation, for here grass grows and water runs. It is rich in thermal and mineral springs, and one of them, the Bartlett, is believed to be a specific for Bright's disease of the kidneys. If tests prove this, Lake County will be sought by many pilgrims. This completes the list of counties which lie in and on the mountain rim of the Sacramento Valley, the other pocket in which the State feels for its money. The counties of that great valley will be considered in a group.

CHAPTER XXVI.

THE SACRAMENTO VALLEY.

THE Sacramento Valley is the State's other pocket Shasta County holds the same relation to this valley that Kern does to the San Joaquin, though it is more mountainous, and its champagne lands are very thoroughly watered. The area of the county is 2,560,000 acres. The Sierra Nevada Mountains and the Coast Range meet in Shasta, as they do in Kern, completing the pocket. This connecting range is lofty, rugged, and well-nigh impassable. Its loftiest peak is grand old Mt. Shasta, bald, snow-crowned, majestic. From Mt. Diablo, which stands at the junction of the two valleys, Mt. Shasta is visible, 265 miles away. Down from its summit, out of the clouds, flows Cloud River, persistently, because mistakenly, called "McCloud" by the California press, and perhaps by the map-makers. Let us do what we can to rescue the name from common-place and restore it to its meaning, expressive of its high descent. Within the county are Lassen Peaks, one of which rises to a height of 10,577 feet. The general elevation of the connection between the two ranges is 5,000 feet, and across this formidable barrier the California and Oregon railway is slowly making its way, in the face of the greatest topographical obstructions that have opposed themselves to railway construction in any part of the continent. When this road has fought its way through tunnels, and by climbing cliffs and clinging to the sides of cañons, it will connect San Francisco with Oregon, and the travel will pass through Shasta and the other Sacramento Valley counties on its way north, toward Alaska, as now, by the Southern Pacific, it goes the length of the San Joaquin, south, toward New Orleans. This Northern route finally connects with a system of railway which is being extended to Hudson's Bay; and there is something suggestive of a great commercial future in the thought that San Francisco will then have railway communication with the Gulf of Mexico on one hand, and Hudson's Bay on the other.

When travel comes from the North over this line, it will, in Shasta, look out upon the first valley land of California, and, fresh from the snowy mountains, it will see here, growing, the orange and lemon tree, untouched by frost, though the latitude is 41° north. Shasta's tillable lands are rich, her climate inviting, the scenery inspiring, and settlers there send out the news of contentment. Next south of Shasta lies Tehama, with an area of 1,958,400 acres, and, spanning the valley from its eastern border, neighbors Plumas and Butte, which lie in the Sierra Nevadas, to the Coast Range on the west. Tehama abounds in streams, and about one-third her soil is that volcanic sort which offers permanent nourishment for the vine. Tehama has obvious natural advantages and beauties, and was the seat of American settlement as early as 1844. Indeed, it attracted the first white settlement north of Sutter's Fort. There is much suggestiveness in the first location made by pioneers when the whole country is open to them, and they can choose at will. Tehama's modern history is vindicating the judgment of her first settlers. The soil is divided between the volcanic and richly alluvial, and there is not much waste in the wide span between the foot-hills of the two ranges.

Irrigation would greatly swell the already fine productions of this county. The facilities for it are so ample that a hydraulic survey demonstrates that one canal taken from the Sacramento River, above Red Bluff, will irrigate 780,000 acres. This, added to the usually good rainfall, would subject every arable acre in the county to high farming and the most perfect productiveness. The crops embrace the full California variety, and there is none greater anywhere in the world.

We have spoken of the Nadeau vineyard as the largest in the world. At Vina, in Tehama County, Governor Stanford has a vineyard which, when extended and completed according to his plans, will take precedence. The old part of this vineyard has produced the phenomenal yield of eleven tons per acre, which, we believe, has never been equaled by vines anywhere else in the world. It is the purpose of Vina vineyard to demonstrate, on a large scale, the ultimate capacity of California vines and vineyard soils, and so far the result is greatly complimentary to the resources of Tehama County. The climate here is clement, as it is throughout the valley, and the scenery varied and agreeable.

Next, down the valley, is Colusa, acreage 1,600,000, with the Sacramento River for its east line and its western border on the top of the Coast Range. Colusa has the honor of having been the scene of the first demonstration of the possibilities of wheat raising; for once it was thought that California would never raise its own bread, and we had to depend on Chili flour. The late Doctor Glenn opened in Colusa the largest wheat farm in the world, 60,000 acres, in one body, and here it was proved that California could bread herself and load fleets with her surplus wheat. So the fame of Colusa went out on the marvelous stories of wheat production, and the county became known throughout the bread-eating world. This county is adapted to, and needs, irrigation joined with drainage. Much of its lands have, at times, too much water, and others have too little. A system of irrigation which necessarily implies the impounding of storm waters will contribute greatly to the needs of each variety of lands. When Colusa County talks of irrigation in the dry season, timid people in Sacramento City feel symptoms of panic at the idea of taking water out of the river. The one real peril which threatens that city is from floods. Her commerce can stand a few weeks of low water; but her foundations cannot stand even a small access to the annual flood which puts her in the center of a great sheet of rather uninteresting water. If Colusa were permitted to impound all the water needed to enrich her arid plains, and make her a principality in productions and wealth, the people of the valley below would not be ague-smitten by the yearly floods, and Sacramento would be delivered from an overflow which has had a sinister effect upon her prosperity and advancement. The county seat is the town of Colusa, and here lives Mr. Will S. Green, editor of the *Sun*, and the most tireless, intelligent, enterprising agitator of irrigation in the State. Pressed by the local interests of his county in the question, he has never ceased to admonish, instruct, implore and beseech the people to be wise in time on this great question.

Sutter and Yuba Counties lie on the east side of the Sacramento River, opposite and below Colusa, and have the general characteristics of their neighbors, while Nevada, Sierra and Placer overlook the valley from the Sierra Nevada Mountains, and have interests kindred to those of its people. Placer is the most noted mountain and foot-hill fruit county in the State, and serves as an example of what can be done in that culture all along the seven hundred miles of foot-hills that lie under the Sierra Nevadas.

Solano, Yolo, and Sacramento lie mostly in the far-gaping mouth of the Sacramento Valley. Solano, once a great wheat county, has come rapidly to the front with her early, medium, and late fruits. Vaca Valley, out of which wheat fields chased the cattle, the harvester replacing the vaquero, has witnessed another transformation, and now is a solid orchard from Vacaville to Putah Creek. Here the cherry, apricot, peach, plum, and grape often ripen earliest, and her lands, that a few years ago were keeping their wheat-raising owners miserable, now bring $1,000 an acre, and the crops taken from them bring a profit that justifies the price. In this region is the Orleans vineyard, in which grow vines from Orleans, France, that were originally from Metternich's Johannisberger vineyard, at Nassau. Here, too, are vines of the "thumb grape," from Italy, and other choice shipping and wine varieties. If a visitor wishes to see fruit farming in its perfection, let him visit Solano County and Vaca Valley. In this county the date palm ripens its fruit unsheltered, and trees grown from the seeds planted thirty years ago may be seen. Yolo County is undergoing the common transformation of her industries, less wheat and more fruit. At Davisville, on the line of the Central Pacific, in this county, is one of the largest raisin vineyards in the State, where may be seen a very fine and successful example of sub-irrigation, the water being discharged from pipes below the surface, thus escaping the loss of evaporation, and always leaving the surface in a condition to work.

Sacramento is the capital county. Its area of 640,000 acres is almost entirely within the great valley, to which it holds the same relative position that Stockton does to the San Joaquin. It is watered by the Sacramento, American, Cosumnes, and Mokelumne Rivers, and has a soil and surface agreeably diversified and very rich. Here Sutter's Fort stood, toward which all immigrants traveled in 1849. Sacramento was first a mass of wagons around the fort, then a city of tents, later of shanties, and now a growing city, with large commerce and manufactures, and destined to have a great commercial future, as the valley above it is developed. The horticulture of this county is well developed, and its agriculture is in the front rank. The city is full of semi-tropical blandishments. On hundreds of private lawns fine orange trees ripen their fruit; the palm, of many varieties, gives an oriental tone to the scene, and flowers in prodigal profusion give the city a Persian air. In this county, on Senator Routier's place, is the most celebrated almond orchard in the State, and so rich, and so varied, and so valuable are the products of the different soils, that if this capital city were limited to the trade, profits, and productions of this one county, it would find in them material for the growth of a metropolis; for Sacramento County alone is capable of producing more wealth from the soil than the whole State of Massachusetts, and yet see the many cities which thrive in Massachusetts!

THE BAY COUNTIES.

Considering the counties that border San Francisco Bay, Napa is the most noted for vineyards. Napa Valley is the modern Eschol, and in the vintage season there, the visitor can hardly believe that he is in America. The scenes and the activities of the season belong to Southern Europe. Here, Krug, Schramm, and many other vintners have developed a wine interest which is of enormous proportions. Napa County is rich in mineral springs and resorts for recreation, and will always be a charming place for the tourist and sojourner, and a place of perfect contentment for the happy people who dwell there.

Alameda, the second county in the State, has for its capital Oakland, the second city of the State, and in many important respects the most strikingly beautiful city on the continent. With one hundred miles of smooth streets, with the bay on one side, and a semi-lune of charming mountains on the other, a system of surpassingly rich valleys behind, and an increasing commerce and manufacturing industry, Oakland offers advantages that are as patent as her charms are irresistible. The agriculture of Alameda is of the first order. The bay climate makes irrigation but moderately necessary, and here are gardens and orchards as fine as the world can boast. In Livermore Valley are thousands of acres of vineyard just coming into bearing, and soon Alameda wines will take their place among the standard varieties of the State.

Below Alameda is Santa Clara County, at the southern arm of the bay, with San Jose, its capital, at the mouth of Santa Clara Valley. This region is unsurpassed in fertility. We would not ask a stranger to believe the truth concerning it, until he sees with his own eyes. It is the opinion of many connoisseurs that the wines of Santa Clara County have reached perfection. Here are made the Naglee brandies which, in a wide competition at Paris, carried away the first premium in a sweepstakes against the world. To the west of the southern arm of the bay lies San Mateo County, which has all the charms, capacities, and resources of the other bay counties. Here, at Menlo Park, are located the homes of millionaires, Senator Stanford, Mr. J. C. Flood, and many others, and here will be that Oxford of the Pacific, the Leland Stanford, Jr., University.

San Francisco City and County are coterminous, and under homogeneous government. A city of 400,000 people, with 200 millionaires, with all the vices and the virtues of a cosmopolitan population drawn from every country under the sun, it is the western Paris, and more nearly resembles ancient Rome, in the possible reach of its influence, derivation of its trade, and diversity of people attracted to it, than does any other modern city. Here meet the Esquimaux from beyond the Arctic Circle, the Klickitat, from Sitka, the Fox Indian, from the Aleutian Archipelago, the Fiji, Samoan, and Maori, the Hawaiian, Malay, Japanese, Chinese, Corean, Hindoo, Parsee, and Arab, the Mexican, Colombian, Greek and Cossack, and people of every nation in Europe. If assimilation of diverse elements makes a strong people, here should be born the future Samson of the nations.

San Francisco is the third commercial city in the Union, and when the means, which we have so often preached in these pages, are found to attract people and make their stay profitable, she will, on this side of the continent, precisely balance New York on the other.

CHAPTER XXVII.

CALIFORNIA—A RESUMÉ.

WE have given this cursory glance at California in detail because it is the oldest State in the New Empire, and in many respects retains the primacy which it so easily gained. Its metropolis is the commercial and financial center of the New Empire, as indeed it is also of the Pacific Islands and coasts. Its products are peculiar to itself. It is the oriental part of America. The seeker will in vain push his quest in Central and South America for a climate and other physical features like those of California. The only climate in the world that approximates it in salubrity, is that of Bermuda, but in its climatology California enjoys the singular distinction of possessing a variety of climates, with subtle distinctions and shades of difference that are perceptible to the invalid in search of health which has been lost in the elemental warfare of the East, and these distinctions and shades of difference are as fixed and reliable as the seasons. From this is derived the advantage of being able to select a location that is curative of a disease, or preventive of an abnormal tendency, in the certainty that its climate is not to change, and force the patient into another exile.

We have specially made plain the value to the commerce of the State that lies in the Sacramento and San Joaquin Valleys. Those valleys through which the San Joaquin River flows north, and the Sacramento south, together, are 400 miles long, by an average of 50 miles wide, and their tillable area of nearly 13,000,000 acres is about one-ninth the total area of the State. If the 20,000 square miles which they cover were as thickly inhabited as New York, the State would have three times its present population. California is larger than Japan, and, permitted to use all needed means to bring its soil into productiveness, it is capable of an annual harvest of fruits, grains, nuts, roots, and berries, as great as the harvest of Japan. Yet California has less than a million of people, and Japan has thirty-five millions, amongst whom poverty is the rare exception. We do not pretend that under our civilization it is possible to crowd a population as densely as it may be done in Asia; but we do say that the natural resources of California are adequate to the task of furnishing a commerce more varied and as great, and to the supplying of food and raiment to a population as large as Japan.

Let us repeat that our references to the counties of California have been kept far within the line of facts, and that in the case of each one referred to there is enough more to be said to make a volume for each individual county. This book is to give to intending immigrants, and the press, a fair idea of the present state of the New Empire, and to stimulate further inquiry on the part of those who may wish to see for themselves the most desirable part of America and the choicest spot on earth.

We have frankly declared the law and the facts upon the irrigation issue. Its importance is under-drawn rather than otherwise, and the frankly declared intentions of the people respecting changes in the law, and, if need be, changes in the courts, have been given as part of current history, and as an admonition to judges and for the information of those who have been deterred from considering a location in California by a judicial blow at the condition of happy and prosperous life in this State. Surely the great rural and business interests of the State are best served by a combination of all agencies, literary and political, to do God's work in forcing even judges to admit that human law is merely a handmaiden to the customary law of Nature, and that the Creator never yet permitted man to mar the pattern, nor destroy the beauty of His footstool.

CHAPTER XXVIII.
IRRIGATION AND DRAINAGE.

HON. A. A. SARGENT, who represented California in the Federal Senate during those years in which the material development of the State required the most congressional help, and got it through his influence, and who later represented his country at Berlin, has, since our chapters on irrigation were written, published in the *Overland Monthly* the following article on this subject. Inasmuch as our say was said without consultation with any legal authorities, we are pleased to find all that we have said indorsed by an authority so great and so trustworthy. By the kind permission of the editor of the *Overland*, we transfer Mr. Sargent's article to these pages, as a conclusive statement of the case:—

"The physical features of California are such that if the law governing the State does not sanction the appropriation of water by diversion to beneficial uses, as opposed to what are called riparian rights, it is a matter of serious regret. It is more. It is a pressing question whether there be not apt and judicious means by which the law may be brought into harmony with public interests.

"The waters of this State are irregularly and scantily supplied by precipitation. Aside from the bays and principal rivers, not available for irrigation, these waters, in their natural state, run through steep, crooked, and rocky cañons to the plains, where they become broad, shifting, shallow streams, often dry, and spread out into swamps, or shallow lakes, the surfaces of which are ordinarily so far below the surrounding country as to be unavailable for reservoirs. The water in these lakes and swamps becomes fetid, fever-breeding, generating swarms of noxious insects, and their neighborhood uninhabitable. These depositories of slimy water can only be drained by intercepting the water which flows toward them in the shallow, scanty streams, which the California vocabulary, for want of a better term, names rivers. These lakes and swamps, to be found in our great plains, are the mere overflow of the streams in the high water of spring, when the snows on the mountains melt with the increasing heat of the sun. At other seasons, they shrink in their beds under excessive evaporation and from absorption, uncovering their dish-like approaches for miles, on which rank tules grow, and rot. The air is poisoned by the exhalations, during the hot season, for miles around; the water turns a light coffee color, and the neighborhood becomes frightfully unwholesome for man and beast.

"If the supply of water from precipitation were greater, and regular, lasting through the year, as in England, the inlets to these lakes would be strong, navigable rivers; the lakes would be deep, clear, and unvarying in size; the swamp would cut out into deep outlets, carrying sparkling waters to the ocean, and freighted with inland commerce. The present nuisances would disappear; for the region about the lakes and swamps would be changed from its natural pestilential condition to salubriousness.

"But we cannot have this greater and regular supply of water from rain and snow. Our climatic conditions forbid it, and will do so for all time. We depend for the little precipitation we get upon the trade winds, which, when conditions, uncertain as the winds, are favorable, send to us in grudging quantities the moisture which tends to make the State habitable. Whatever the direction of the wind in England, it traverses high seas ordinarily in commotion from storms. The isle is drenched at all seasons, except occasionally in some of the spring months, and it is liable during those months to rain enough to make a feature in a California 'wet season.'

"In consequence of this feature of our climate (so

strongly in contrast with that of England, whence comes the doctrine of riparian rights), when our people began to settle the valleys of this State, they found these swamps and swampy lakes, which Nature had already fashioned. In the progress of settlement, the question has arisen, Is it necessary or right to keep forever these polluting areas, or can engineering science and public necessity obviate them? The soil under the thin layer of water in the lakes is rich; may it be made cultivable? Homes may be made in the region now too unhealthy for any population; shall the State be allowed to improve? Enterprise stands ready to create taxable property there; is there a law which forbids it?

"The method of redemption is plain, if there is not something in the law to prevent its use. We repeat, that the problem is to get the water out of depressions in the valleys too low for ordinary drainage. Thus, Tulare Lake is the overflow from King's River. It has no outlet, unless the overflow from King's River becomes so great that the surface of the lake rises high enough to send the water back through the same river to an outlet in the San Joaquin. Kern and Buena Vista Lakes are the overflow from Kern River, a sumpage ground in spring. The water, when there is enough of it, flows into Buena Vista Slough, and thence south into the lakes. If the lakes get full enough, the water flows back through Buena Vista Slough to Buena Vista Swamp, where it is spread out and lost. Under such conditions, it is obvious that there can be no drainage of Tulare Lake through King's River, or of Kern or Buena Vista Lakes through the Buena Vista Slough or Swamp. The great body of the water is condemned to fester and dry up in the hot sun of that region, with the effects described. But the waters of King's River, on their way to Tulare Lake, and those of Kern River, on their way to Kern and Buena Vista Lakes, may be intercepted and the water be used for irrigation on the parched plains; and then the lakes and swamps will permanently dry up, their beds be given to fertility and man, their noxious insects disappear, their fevers vanish, and prosperity take the place of desolation.

"This is one side of the problem; but there is another and more important one. The great valley of California lies between latitude 34° 50' near Fort Tejon, and 40° 40' near Shasta, giving an extreme length of 450 miles, and an average width of 40 miles, including the foot-hills. It lies between the Coast Range and the Sierra Nevadas, and within the cup of the mountains lies an area of 52,200 square miles, equal to half of all the Middle States. In this great valley are millions of acres of land, possessing all the elements of fertility except moisture, a climate agreeable in winter, hot and desiccating in summer, and yet not enervating nor unfavorable to industry. Under the stimulus of water, from 50 to 80 bushels of wheat per acre have been produced, and 45 bushels of barley as a volunteer crop. Five crops of alfalfa have been grown in one year, yielding an average of 15 tons per acre. From the farthest bound of the Colorado desert to the headwaters of the Sacramento, is a region to be benefited by irrigation; and one-half of it, approximating in fertility to that above described, is absolutely sterile without it. This part lies, year after year, as it has done since the mountains took their present form, dreary, dead, and forbidding, except in comparatively limited areas, where a system of irrigation has been adopted, changing sightless deserts into scenes of perennial loveliness. The traveler through Tulare, Fresno, Kern, Stanislaus, Merced, Los Angeles, and other southern counties, may see, lying side by side, desert tracts parched and burnt like the Sahara, and oases of wondrous beauty, whereon tropical fruits flourish in the vicinity of grain crops; where rich meadows feed innumerable herds of cattle, horses, and sheep. A few years ago the oasis was desert. What magician has changed this so much for the better? Redemption was effected by bringing the fugitive and scanty water of these streams to these lands, and thus quickening them into life. As the area of irrigated land has extended, all branches of business have become enlarged.

"A great wool clip, raisins, wine, and brandy, oranges and other tropical fruits, countless herds of cattle fattened for home and foreign markets, growing villages and cities, pleasant and numerous homes, all attest the benefits accruing from irrigation. A great trade has sprung up, to the advantage of Los Angeles, San Francisco, and the whole State. This is the result from the irrigation of a few hundred thousand acres of land, to water which ditches and canals thousands of miles in length have been constructed and maintained at a cost of some $100,000,000. It is calculated by engineers that the water of our rivers, available for the purpose, will be sufficient to extend irrigation to many millions of acres which now are absolutely useless, but which will then be as fertile as the Nile Valley after a swelling of the sacred river.

"These facts show the relation of irrigation, or the appropriation of water to useful purposes, to the problem of draining the pestilential lakes and marshes of the State. By constructing reservoirs in the mountains to catch the surplus water, and by intercepting the water on its way to the stagnant pools which naturally receive it, and where it is wasted by evaporation, and by spreading it out over desert lands, the swampy lakes and morasses are dried up, and become the scenes of agricultural prosperity, while thriving farms are created on the deserts to embellish and enrich the State. Works of irrigation and for reclaiming marsh lands go together in all old countries where either are needful.

"If it be true that the Legislature has been so improvident in its laws that the people of the State are powerless to dry up their swamps and fertilize their deserts, then the population of the State is too large, and its prosperity is built on so insecure a basis that a collapse is impending. If this be true, the colonies of Fresno, Anaheim, Riverside, etc., have chosen the wrong State for their settlements. The farmers who have created cultivable land in Tulare Lake must soon see their possessions engulfed in the returning waters. The prosperous farms in the deserts must return to their original sand heaps; the verdant crops that beautify a broad region must die, and the herds that feed there must die with them, or be driven away. Towns must dwindle to villages, and villages and homesteads disappear. All industries built upon irrigation must perish when irrigation ceases, and future improvements conditioned upon irrigation be denied. These propositions are so simple that they are axiomatic. They are founded in the experience of all arid countries. All our libraries contain shelves full of books illustrating them.

"It may well be supposed that this people will not submit to such consequences without an earnest attempt to avert them. It can hardly be anticipated that they will accept the destruction of such solid interests upon the fiat of four Supreme Judges, when three other members of the same respectable tribunal dissent, and say the majority is mistaken in its law. By our form of government, there is an appeal to the people from all executive or judicial action. By making the judiciary elective, the Constitution devolves the duty upon the people of determining as to the fitness of judges, and makes these directly responsible to the people. Many oldschool thinkers have objected to this feature of modern constitutions; but it has survived all attacks, and is now firmly rooted in public policy. By that policy the people have opportunity to confirm or reverse the decisions of their judges, and may reasonably be expected to exercise this power in a case where public interests are put at hazard, and the decision of the court meets with general popular non-concurrence.

"The effects of the decision in question are not localized to the great valleys of the State. The mountains are seamed with water ditches, constructed at immense cost for mining purposes, in defiance of riparian rights. Some of these canals are already utilized for irrigation, and more will be in the future, if it is permissible. For this purpose they need to be greatly extended, and new ditches to be taken out below the present points of diversion. Is the miner, driven from his occupation by the action of courts, to be prevented by the courts from maintaining his means of diversion, or creating new ones, to fertilize the vineyards and orchards he is planting in the foot-hills? The few dwellers along the rocky cañons are the riparian proprietors, and they are the ones who can compel the appropriators to turn the water back into the streams, that it may run unused by their solitary cabins.

"The question, therefore, whether what has been heretofore held as the common law of California—viz: the right of the first appropriator of water for beneficial use, to enjoy it to the extent of his appropriation—or whether recognition as conclusive of the inapplicable common law of England (which gives to the proprietor on the banks of a water course the right to have all water naturally flowing by or through his land, continue so to flow, unused, undisturbed and undiminished) shall prevail, becomes a vital one to all the people of this State to consider, both in economic and legal aspects.

"By the act of April 13, 1850, the California Legislature enacted that 'The common law of England, so far as it is not repugnant to, or inconsistent with, the Constitution or laws of the State of California, shall be the rule of decision in the courts of this State.' Upon this enactment, the structure of 'riparian rights' rests, and the right of appropriation is denied, however destructive the consequences. The State then signed the bond, giving the pound of flesh; it enacted away all control, ownership, and beneficial use of its waters, and improvidently wrote ruin upon most of its territory. So runs the argument. It is necessary to its conclusiveness to insist that the common law is inflexible in its provisions, unbending to circumstances, uninfluenced by the necessities of the people, which its provisions govern. The laws of legislatures may be changed, constitutions be modified by amendment or explained away by courts; but the common law of England is fastened on the State, and may throttle it, and there is no relief, unless judges in England vary its tendencies. New conditions may arise here; but they must yield to it. New discoveries may be made in art, science, and political economy, of all of which the originators of the common law had no conception; yet they must wait upon its teachings and abide by its slightest indications. No people ever assumed meekly a more intolerable yoke, or submitted to a more absurd bondage, if this be true. But it is not true. One of the leading principles of the English common law is, that it is flexible, and may be modified to suit the varying wants of the community. Were this otherwise, it would never have been taken by English colonists to their new homes. The declaration of rights made by the first Continental Congress in 1774, declared that 'the respective colonies are entitled to the common law of England, and to the benefit of such English statutes as existed at the time of their colonization, and which they have by experience found to be applicable to their social, local, and other circumstances.' Unless so applicable, the common law was repudiated by the Continental Congress, as England would have repudiated it if it had ceased to be applicable to her necessities.

"The United States Supreme Court has declared that the common law of America is not to be taken in all respects to be that of England, but that the settlers adopted only that portion which was applicable to their situation. The constitutions of many States contain language similar to the statute of 1850, and contain no words of exemption of such portions of the common law as are inapplicable to the condition or necessities of the particular community; notably in New York, New Hampshire, and Massachusetts; and yet the courts in those States have held that the common law is not a rule of decision where opposed to the wants of the people.

"As an illustration of the modification of the English common law in the United States may be instanced the case of ancient lights. Blackstone says: 'If one obstructs another's ancient windows, the law will animadvert hereon as an injury, and protect the injured party in his possession.' This doctrine is as well seated in the English common law as is that of riparian rights. Any one passing Cheapside and other busy traffic streets in East London, will see where, in the march of modern improvements, old buildings have been pulled down to erect finer structures. On the squatty neighboring buildings, at little windows looking out on old courts or alleys, are put numerous signs bearing the inscription, 'Ancient Lights,' as a warning to the neighbor not to build his new building so high or in such shape as to obstruct the light through these old peep-holes. The same author defines the common law to be general customs which are the universal rule of the whole kingdom, and are ascertained and their validity determined by the judges of the several courts of justice. This common law, he says, protects these ancient lights. The lead in repudiating the common law doctrine of ancient lights in the United States, was taken by the courts of New York, the Constitution of which State makes the common law the rule of decision to the extent to which it is so made by our statute. Upon a case calling for a decision as to the right to obstruct an ancient light, the learned judge repudiated the English common law doctrine, upon the ground that 'it cannot be applied in the growing cities and villages of this country, without working the most mischievous consequences. It has never, we think, been deemed a part of our law.' The same ruling has been made by every court in the United States, save one, which has passed on the question. Yet this doctrine is incrusted in the common law, as every lawyer knows. Even so the doctrine of riparian rights cannot be applied to our arid State without the most mischievous consequences. Why, then, apply it?

"The same great jurist said: 'I think no doctrine better settled than that such portions of the law of England as are not adapted to our own condition, form no part of the law of this State. The exception includes not only such laws as are inconsistent with the spirit of our institutions, but such as were framed with special reference to the physical condition of a country differing widely from our own. It is contrary to the spirit of the common law, to apply a rule founded on a particular reason, to a case where that reason utterly fails.'

"The doctrine of riparian rights grew up in a small country, continually drenched with water, where the necessity for irrigation was unknown, and the only use of water was for navigation by shallow boats, or to propel water mills. In England the annual rainfall reaches eighty inches; in some parts of California it does not exceed six inches. The problem in England has always been to get rid of water, not to divert it, for there was no beneficial use for the diverted waters. But the doctrine of riparianism grew up anciently, when the owners of grain mills along the streams desired the water to flow steadily to the rude mill wheels, and the movers of country products, before railroad transportation, desired to prevent obstructions being put in their way in the streams. The judges moulded their decisions upon these narrow necessities, and on kindred ones in the course of time. Their doctrines fitted the times and the necessities of the communities to which they applied. They are out of place in an arid region, where navigation of streams available for irrigation is impossible, and the fluctuating supply of water precludes its use for power. As the common law was devised to minister to the wants of the community governed by it, and enable them to make the most of their surroundings, it is obvious that the judges would have sanctioned appropriation for irrigation, had irrigation been a great necessity for England. To doubt this is to misunderstand the mode of growth of the common law. The rule of riparianism was founded upon particular reasons. If the reasons had been different, the rule would have been different also. It is therefore a violence to good judgment to import into our law a rule founded on reasons which have no existence with us; indeed, where the reasons are exactly opposite. To do so is to violate the common law, not to enforce it. The writer entertains the highest respect for Hon. Allen Thurman as a statesman and jurist, and such is generally conceded to him. Judge Thurman, when upon the Supreme bench of Ohio, laid down this principle in plain language. He said: 'The English common law, so far as it is reasonable in itself, suitable to the condition and business of our people, and consistent, with the letter and spirit of our Federal and State constitutions and statutes, has been and is followed by our courts, and may be said to constitute a part of the common law of Ohio. But whenever it has been found wanting in any of these requisites, our courts have not hesitated to modify it to suit our circumstances; or, if necessary, to wholly depart from it.' Would space permit, it might be shown by a wide range of quotations from eminent judges and law writers, that the common law of England is not enforced in American courts where such application is not consonant with our condition and necessities. Our Supreme Court had abundant precedents and the highest authority to decide, if it so willed to decide, that the doctrine of riparian rights, originating under circumstances and for reasons so different from those existing here, is not the law of this State, and never has been.

"An illustration is furnished by the courts of setting aside the English common law, because the physical conditions of this country are different from those of England, as regards admiralty jurisdiction. The early decisions of the United States Supreme Court assumed that the expression 'cases of admiralty and maritime jurisdiction' was used in the Constitution in the same sense as in England at the time the Constitution was framed; and therefore, following the restriction which the common law had imposed on admiralty in England, held that the jurisdiction was limited to matters on the high seas or tide waters, and not within the body of a country. The earlier cases adopted the language of the law of England, where the navigable waters are tidal; but the same court afterwards held, and still holds, the rule inapplicable in this country, which has great inland seas and long public rivers, navigable to long distances beyond the set of the tide. It recognized 'the necessities of commerce,' as requiring the application of the jurisdiction to all public navigable waters on which commerce is carried between different States or nations. Yet the same great tribunal has always held that the English common law, where our conditions permit its useful application, is the heritage of the people of this country; that is, that it is a minister to our prosperity, and not a drag upon it. The Act of 1850 did not, therefore, upon the principles of construction applied by other jurists, import the doctrine of riparian rights into this State. Had it been intended so to do, surely no legislature ever so little understood, or was so careless of, the heritage of its constituents and that of their children's children.

"That legislature met six months before the State itself had a legal existence. It was made up partly of natives who knew nothing of irrigation, who only valued land for pasturage, and watered their herds at any convenient spring. If they understood Mexican laws with regard to water, which is doubtful, they knew that this was subject to common use, and could be kept in the natural channel, or diverted by individuals or corporations, as the Government permitted. Riparianism was unknown to them. The remainder of those legislators were gold-seekers or office-hunters, who necessarily had ittle knowledge of the physical geography of the State, and hence were poorly qualified to pass an intelligent judgment on this question, even if they gave it a thought, which there is no evidence that they did, and which they undoubtedly did not. They resorted to the mines from the halls of legislation, and aided to establish a custom of appropriation for mining purposes, which was illegal under the modern construction of their innocently adopted statute. But it is important that under the decisions of the Supreme Court in its early years, this system of appropriation of water for mining purposes grew up, and was recognized as legal. The judges who made those decisions were near the period of enactment, and their views have the value of contemporary construction. The policy which they sanctioned was afterwards reviewed by the United States Supreme Court, and that court said: 'As respects the use of water for mining purposes, the doctrines of the common law declaratory of the rights of riparian owners were, at an early day after the discovery of gold, found to be inapplicable, or applicable only in a very limited extent, to the necessity of miners, and inadequate to their protection;' and in another case the same court said that the views expressed, and the rulings made, in regard to the appropriation of water for mining purposes 'are equally applicable to the use of water on the public lands for purposes of irrigation. No distinction is made in those States and Territories by the customs of miners and settlers, or by the courts, in the rights of the first appropriator from the use made of water, if the use be a beneficial one.'

"That tribunal recognized that the customs and necessities of the people of this coast had moulded a common law for them in this particular, and that the common law of England was inapplicable and mischievous, in that it was, as they said, 'incompatible with any extended diversion of water, and its conveyance to points from which it could not be restored to the stream.'

"Colorado has put into its Constitution a provision recognizing the priority of right to water by priority of appropriation. Like California, it is arid, and needs irrigation to fertilize its fields. It has already solved this question, as it was proposed by the recent State Irrigator's Convention to solve it, by organic law. But the Supreme Court of that State had decided that the doctrine of riparian rights had no applicability to Colorado even before the adoption of the constitutional provision; because imperative necessity, unknown to the countries in which the common law originated, compelled Colorado to recognize appropriation. The reasoning of that court is so just, its recognition of the great necessities of the State so clear, and the parallel of circumstances with those of California so exact, that it is well to cite the decision at some length:—

"'It is contended that the common law principles of riparian proprietorship prevailed in Colorado until 1876, and that the doctrine of priority of right to water by priority of appropriation thereof was first recognized and adopted in the Constitution. But we think the latter doctrine has existed since the date of the earliest appropriations of water within the boundaries of the State. The climate is dry, and the soil, when moistened only by the usual rainfall, is arid and unproductive; except in a few favored sections, artificial irrigation for agriculture is an absolute necessity. Water in the various streams thus acquires a value unknown in moister climates. Instead of being a mere incident to the soil, it rises, when appropriated, to the dignity of a distinct usufructuary estate, or right of property. It has always been the policy of the National, as well as the Territorial and State Governments, to encourage the diversion and use of water for agriculture; and vast expenditures of time and money have been made in reclaiming and fertilizing by irrigation portions of our unproductive territory. Houses have been built, permanent improvements made, the soil has been cultivated, and thousands of acres have been rendered immensely valuable, with the understanding that appropriations of water would be protected. Deny the doctrine of priority, or superiority of right by priority of appropriation, and a great part of the value of all this property is at once destroyed. . . . We conclude, then, that the common-law doctrine, giving the riparian owner a right to the flow of water in its natural channel upon and over his lands, even though he makes no beneficial use thereof, is inapplicable to Colorado. Imperative necessity, unknown to the countries which gave it birth, compels the recognition of another doctrine in conflict therewith.'

"It is a matter of extreme regret that a few more of the members of our own Supreme Court could not see judicially, or give due weight to, what the Supreme Court in Colorado so clearly sees and applies; viz., that it has been the policy of the National and State governments to encourage the diversion and use of water for agriculture; that vast expenditures of time and money have been made in reclaiming and fertilizing by irrigation, portions of our unproductive territory; that houses and villages have been built, costly, permanent improvements made, and hundreds of thousands of acres rendered immensely valuable, which else would have remained desert, with the understanding that appropriations of water would be protected, and that the denial of the right of appropriation destroys this vast property.

"The Supreme Court of Nevada, in an early case, sanctioned the doctrine of riparian rights. But it has since retreated from that ground, and approved the doctrine of appropriation, holding that priority of appropriation is a test of superiority of right. Its views as to the great system of antiriparianism, built up in the early days in this State, and the sanction it received from the courts, are expressed as follows: 'In all the Pacific Coast States and Territories . . . the doctrines of the common law, declaratory of the rights of riparian proprietors respecting the use of running waters, were held to be inapplicable, or applicable to only a very limited extent, to the wants and necessities of the people, whether engaged in mining, agricultural, or other pursuits; and it was decided that prior appropriation gave the better right to the use of the running waters to the extent, in quantity and quality, necessary for the use to which the waters were applied. This was the universal custom of the Coast, sanctioned by the laws and decisions of the courts in the respective States and Territories, and approved and followed by the Supreme

Court of the United States.' It may therefore be said, on the testimony of this Supreme Court of Nevada, and the Supreme Court of the United States, that there is a universal custom or common law established in this State by the concurrence of miners, farmers, and courts, by which appropriation was established and riparianism rejected as the law of this State.

"In the view of our high court, there is no public policy which can empower it to disregard or modify the common law of England because of a benefit to many persons; and it holds it doubtful, if it is to the common benefit, or the benefit of many persons, to promote the appropriation of water for agricultural purposes. Upon the latter proposition, the people need no decision; they are as nearly unanimous as possible that the court is all wrong. As to the first proposition, if the principle of it had been adopted by the Supreme Courts of New York, Massachusetts, Maine, Pennsylvania, Ohio, Texas, Illinois, etc., the common-law doctrine of easement in ancient lights would be the law of this country, and such structures as the Nevada Block or Safe Deposit building, and the many palaces of trade in our growing cities, could not have been built without the payment of enormous sums of smart money; or, as the court puts it, 'on payment of due compensation.' But these courts, and others, recognized the argument *ab inconvenienti*, and enforced it. Did they not 'legislate in such manner as to deny citizens their vested rights'? Our Supreme Court would so characterize this action, and it refrains from imitating the example of most of the Supreme Courts of the Union in a parallel case. It is held more strictly by the tether of the common-law than the other courts are. It cites authorities from those courts to justify its adherence to the common law upon riparian rights, but underrates the example of the same courts where they depart from the common law because the reason for the law fails in their communities. But there are wide climatic differences between California and the States in question. West of the one-hundredth meridian, the country is arid; east of it, the climate approximates to that of England, and irrigation ditches are almost unknown. Regular rains, distributed through the season, obviate costly works for diversion and distribution of water, and leave no room for dissent from the English doctrine of riparian rights. Hence the courts follow the common law in that regard. They have no reason to do otherwise. What they will do where they find the common law 'not adapted to the necessities of our growing communities,' they have shown. Those illustrious judges would have undoubtedly as freely decided that the common-law doctrine of riparian rights is on a level with the common-law doctrine of ancient lights, if they had lived in a country whose prosperity depended upon diversion and irrigation; and that it could as little stand in the way of progress and civilization. We must have a common law for the region west of the one-hundredth meridian, and courts which can see its necessity, and enforce it. An eminent law writer (Wharton) has discussed the proposition whether judges can or should legislate: 'Judges are not legislators for the purpose of revolutionizing the law, but they are legislators for the purpose of evolving from it rules which should properly govern present issues, and winnowing from it limitations which are withered and dead. And when this duty—a duty which is a necessary incident of judicial office—is frankly recognized by the judiciary, the process of legal development and of suppression will be carried on more effectively and wisely than it can be done by those who shut their eyes to the duty. For no disclaimer can relieve the judiciary from the function of gradually modifying the law, by adoption and rejection.'

"It may be respectfully suggested that our Supreme Court fails to carry its premises to their logical conclusion. It holds that the common law of England was adopted in this State, and that in that law riparian rights are entrenched. The common law upon riparian rights is substantially as follows:—

"'Every proprietor of lands on the bank of a stream has an equal right to use the waters which flow in the stream, and consequently no proprietor can have the right to use the water to the prejudice of any other proprietor. Without the consent of the other proprietors, no proprietor can either diminish the quantity of water which would otherwise descend to the proprietors below, or throw the water back upon the proprietors above.'

"The Supreme Court dispense with this rule of the common law, in favor of a supraparian proprietor, by holding that he may use on the land at the head of his ditch, any reasonable quantity of water for irrigation, if he return the surplus to the stream. Suppose there is no surplus? But this scanty privilege is a modification of the common law, and not the original doctrine. It was not the common law in 1850. Since that date, certain judges of England have expressed some hesitating assent to 'the American doctrine of appropriation,' in the case of supraparian proprietors; and hence a California court ventures also to give it a qualified assent. Are we, then, governed by the House of Lords in England, not by our own legislature and courts? The English courts are daily making laws adapted to their country, and thus our judges wait to apply them to ours. There should be law quotations telegraphed, like stock quotations, or the price of wheat. It would be strange if, in all the dictum and rubbish spoken by innumerable courts, there could not be found some warrant for this subservience to foreign tribunals; nevertheless, the better, safer, and more dignified rule would seem to be that laid down by an eminent law writer (Sedgwick), who says:—

"'It has been uniformly adjudged in this country, that the common law, however adopted, is in force here *only so far as it is adapted to our situation, wants, and institutions.*'

"To refuse to apply it where it is opposed to our situation, wants, and institutions, is not to legislate; it is only to discriminate.

"The common law was not adopted in this State, or any other, as a code, but as a 'rule of decision.' It is not compulsory, but advisory. It is useful only where it is reasonable. It depends for its applicability upon the soundness of the reasons supporting it, and the similarity of the conditions in given cases. It certainly stands upon no firmer footing with us than in England, and there judges daily enlarge, contract, or explain it away.

"The recent advance in the English courts towards appropriation of water is an illustration of the flexibility of the common law, and their mode of treating it. As long as the only use for water was to float craft, or drive machinery, they adhered to the stricter doctrine. But of late years the use of flooding has become partially understood in the west and south of England, to increase the produce of grass by converting the land into water meadows. Poor heaths have been converted into luxuriant pastures by the use of irrigation alone. Quick to detect changes in public wants, the courts have recognized this additional use of water; but, as every water course has an owner, and only the owners seek to divert its water, the decisions have not advanced farther than to favor, in some degree, the claims of riparian appropriators to beneficial use. Upon the strength of such intimations we also advance a short step, not venturing to go alone, or to do what the same English courts would do in a proper case—set aside all previous adjudications to serve the public interests, as did the United States Supreme Court in the matter of admiralty jurisdiction, and our courts generally in the case of ancient lights.

"A disheartening portion of the opinion of the majority of our court, is that wherein they undervalue the benefits that have been gained under the appropriation system, and discredit those of the future. With such impressions upon that vital subject, it was easier to decree practically that irrigation in this State shall be confined to narrow margins along water courses, and that the great plains beyond shall rest in perpetual barrenness. If an outlet of escape from this condition was left open, by condemning upon compensation all the available waters of the State, it is through a course of expense so frightfully great that no sane man can expect to see it realized. The day that decision was rendered, running water, to which there had hardly been a claimant except the industrious appropriators, became worthless to them, and worth hundeds of millions of blackmail to loiterers. Such counties as Fresno, Merced, Stanislaus, Kern, Tulare, Los Angeles, and San Bernardino, such towns as

Fresno, Bakersfield, Riverside, Pasadena, etc., received a staggering blow, from which they can recover only by a return to what was before believed to be the policy of the law. The curse of disputed land title is not worse than that of disputed water rights; and where water is a condition of existence, as in the region named, the curse is fearfully aggravated. On that day a hundred million dollars invested in irrigation ditches, and thrice that amount of improved farms, orchards, and vineyards, became the sport of litigation, with the disadvantage of prejudgment.

"The decision was made in a case not necessarily calling for it. The plaintiffs claimed under a grant of swamp lands from the State, the condition of the grant being that they should free the land from water by draining it; or by turning the water away from it. But the plaintiffs claimed the right to have all the water flow to these lands that would, in the course of nature, flow there; in other words, they held the land on condition of making it dry, and they claimed the water to keep it wet. Again, the decision deals with, and virtually denies, the right of the defendants to divert the waters of Kern River for irrigation purposes, because, say the court, the plaintiffs are riparian proprietors, not on Kern River, but in a swamp that is made by the chance overflow of certain lakes, which are not a part of that river. The question has been asked why, in a matter of so much moment as that of laying down a rule of property affecting so seriously all the business interests of the State, the court did not wait, as requested, until a case arose where the facts demanded it.

"As the water that reaches the plaintiff's swamp lands is that only which overflows during the brief period of melting snows from Buena Vista and Kern Lakes, it necessarily follows that these lakes must be maintained to keep the swamp lands so supplied. Professor George Davidson, in his report upon irrigation in California, speaks of these lakes as he found them as early in the year as May, as lying in a temperature of 130°, and being 'very green, warm, and unfit for domestic use.' This enormous heat, and the cessation, so early in the spring, of water supply from the mountains, 'causes a large area of land,' says another observer, 'to become alternately wet and dry, producing a great mass of vegetation, the decay of which causes a good deal of malaria, carrying sickness over a wide region, and as far as Bakersfield. Enormous swarms of mosquitoes are generated, which infest the swamp and lakes, stinging cattle and horses to madness, not only around the lake, but at long distances from it. Cattle drinking the water, or feeding at the lake, are sickened by fevers, and the lake becomes a most annoying and deadly nuisance. It is a sheet of ever-varying, stagnant water, good for nothing but producing malaria and mosquitoes. Even the fish propagated in its waters are not fit to eat.'

"The direct effect of the decision is to perpetuate this great nuisance, which the police power of the State should be employed to abate. But this is of less consequence, as, if the great system of reclamation by irrigation inaugurated in the southern valley is to be stopped, it matters not whether the air in the solitudes so enforced is poisonous or not. They will necessarily relapse to their desolate condition of twenty years ago, when the traveler passed over fifty miles at a stretch without finding a human habitation. Under the system of riparianism, as expounded by our judges, the great plains will again become, as they were for the first twenty years of the State's existence, habitable only by wild hogs and gophers. The lakes and morasses may therefore be allowed to remain, to yield their fragrant tribute to the English common law.

"The artificial and fragmentary way in which great questions are sometimes tried in courts prevents a large consideration of them. It may be insisted that in this case, under the issues, all these considerations were not, and could not be, urged. Yet, under all disadvantages, it could not be overlooked, even if underrated, that one side of the question represented the reclamation of our broad deserts and of these swamps, the health of the community, its prosperity in the largest sense, and the creation of productive property. On the other was a policy that would keep these lakes full of stagnant water, compel the overflow of Kern River to find a perpetual deposit there, destroy the health of the region, infest it with intolerable pests, condemn the uplands to sterility, and break up inestimable industries. Every farmer in the great valleys was interested in the decision of the question, for all live by irrigation. Every dweller in farm houses near these and other such lakes, and in the surrounding villages, had a vital interest to know if miasmatic air should steal, under the protection of the law, into his home at night. The merchants and manufacturers of the State had an interest in its decision; for if the farmer was ruined, he could not buy or pay. All who desired the State to be developed, its vast arid plains to yield the abundance of which, under conditions, they are capable, were interested in it. It is in the view of this wide and absorbing interest of the whole State that this discussion of the facts and principles involved is attempted. The personal aspects of this *cause célèbre*, however important to the litigants, sink into insignificance compared with the great interest of the State in the ultimate determination of the question whether the means which, as we shall see, all countries physically conditioned like ours have employed to promote their growth and happiness can be permitted in this State, or shall be denied because countries differently circumstanced have never felt the need of, or employed them. The system of appropriation is not hostile to the real interests of the riparian proprietor, provided he will avail himself of its advantages. It is inconsistent with the practice of wasting the waters of the State by letting them run idly into the unthankful ocean; but it is not inconsistent with the use of water by any one, riparian proprietor or not, who will take the necessary steps to appropriate and put the water to some beneficial use. Nearly all riparian proprietors are appropriators in the sense here intended. They have put up their notices claiming water, and dug their ditches leading to their irrigated fields, or to tanks for stock. The decision of the Supreme Court is hurtful to such proprietors in most cases; for they need to irrigate over wider spaces more liberally than the limiting words of the court permit. Such riparian appropriators are injured by the new departure in law, as much as any other. Water is so precious in this State that every means must be used to husband it. Every drop that falls into the sea has failed of its mission. In the Coast Range, where thin threads of water run, and are apt to dry up, or sink away in the hot summers, it would be well to imitate the example of the old padres, who concreted the beds of the little streams, or made concrete ditches along their banks. This preserves the water, and it is appropriation as well.

" The doctrine of riparian ownership will be very difficult of application in this State, for other physical reasons than those existing in its climate. All the streams of Southern California, after they leave their rocky cañon beds, run through shifting sands. In many cases they have no defined banks, or steady course, but shift their direction under the effect of storms. These shifting streams break away, during high water, from their temporary beds, and take new courses, often widely diverging from previous ones. The river affected by this suit will illustrate. In 1862 it ran below where Bakersfield now stands, southeasterly, and discharged into the east end of Kern Lake, when there was water enough to get through the sands so far. In 1867 it changed, during a storm, to what is now called Old River, and discharged through one fork at the west end of the lake, and through another still farther west into the slough connecting that lake with Buena Vista Lake. It now runs still farther west in New River, and discharges northwest of Buena Vista Lake into Buena Vista Slough, whence it drops back, southerly, to the lake, in an opposite direction from Buena Vista Swamp. The point of discharge, in each case, is about ten miles from the previous one. The original United States surveys, made in 1855, show a still wider divergence of this shifting channel. Such rivers refuse to be governed by the decrees of courts that 'inseparably annex them to the soil, not as an easement or appurtenance, but as part and parcel of it.' An appropriator easily adapts his means of diversion to such streams; but a riparian proprietor finds his inseparable annex nearly as fleeting as the clouds that sail over his land. In whatever light the matter is viewed, the conclusion comes irresistibly back, that the laws made for a country so different in all physical aspects as England is from California, cannot, and ought not to be enforced here.

"In the foreign possessions of England, the practice of appropriation prevails over the doctrine of riparian rights, wherever irrigation is a necessity. It is so in India and in Australia. India has gigantic works for systematic irrigation. Three hundred and seventy millions of British money are being expended in that country to supplement a system older than our era. Professor George Davidson reports that the whole breadth of the base of the peninsula of India, sweeping in a great curve from the delta of the Ganges to the delta of the Indus, is the field of a vast system of irrigation. The supply of water is in the Himalayas, where snows ensure an unceasing supply. The Rocky and Sierra Nevada Mountains are the Himalayas of the arid region of the United States, while the broad areas of irrigable lands which adjoin them are, perhaps, equal in extent to the great plains of India. For over two thousand years the people of India have cultivated by means of canals and reservoirs, and English capital has projected and commenced great works, with better engineering science and wider reach. The effects are already seen in the world's markets by the competition of the wheat and cotton of India. The rains of India are usually confined to a single month. Though copious for that period, they do not give the continued moisture necessary for crops. In the densely populated parts of the country, two crops annually are necessary to feed the people, and these can be had only by utilizing, by irrigation, the water caused by the melting snows stored in the mountains. The alternative of less production is starvation, with the attendant fevers. The director of the Ganges Canal Water Works states, as a striking advantage of irrigation in that country, the substitution of a constant for a fluctuating return of produce. Alternations of production and failure consequent upon non-irrigable agriculture, are significant of enormous misery among the laboring classes. These have disappeared as the great works inaugurated by English capitalists have become operative. In a community dependent for its means of subsistence on the soil, the importance of having thus excluded the disturbing influence of variable seasons need not be insisted on. All the benefits of security for capital invested in cultivation are obtained; the revenue fluctuates only with the price of produce, and the working classes have cheap food, and a constant demand for their labor. The horrible famines of India, the sickening details of which have, from time to time, reached our distant ears, cease where irrigation gives steady returns to the labors of the husbandman. In India the Government possesses the right of property in all running waters whatsoever. It may dispose of them forever, if it thinks fit, and the doctrine of riparian rights has no part in the economy of that country.

"Irrigation is resorted to in all countries where much of the land must otherwise remain barren from drought. In Egypt it was practiced two thousand year before Christ, by means of great canals and artificial lakes. Extensive works, intended for the irrigation of large districts, existed in times of remote antiquity, in Persia, China, and other parts of the East, and such works still exist, and provide food for the teeming millions who would else perish. Irrigation is a powerful agent in the plains of northern Italy, and the Government recognizes its economic importance, encourages it by every means, and is especially careful in the education of civil engineers, the highest grade among whom is the hydraulic engineer. The length of canals in Lombardy alone, is over five thousand miles, and there is scarcely an acre of the Milanese that is without several intersecting canals. In round numbers there are a million acres irrigated in Lombardy. The system has been perfecting for seven hundred years, and has gone on under all changes of dynasty and all civil commotions. It has converted a barren waste into a garden. The right of property in all running waters, whether of rivers, steams, or torrents, appertains to the Government. While the

Government disposes of the waters of all rivers and canals, it recognizes the claims of towns, or associations of proprietors, to the supplies which they have enjoyed by prescriptive title for long periods of time. Private rights to divert water have grown up to such extent that the right asserted by the State, is nearly a barren one, and its enforcement has reference rather to administration and police duties than to direct financial considerations. In exercising its right of property in waters available for irrigation, the Government of Lombardy follows one of three courses. First, it disposes of the water in absolute property, to parties paying certain established sums for the right to divert it. Second, it grants perpetual leases of the water on payment of a certain annual amount. Third, it grants a temporary lease for a variable time at a certain annual rate, the water reverting to the State on the termination of the lease. By far the most common of these courses is the first, and it operates the most beneficially. The origin of the system of irrigation was with the great landed proprietors upon their properties. With the revival of knowledge in Italy, the art of hydraulic engineering was called into existence, and the extensive demand for skill in its details created, early, a supply of men familiar with all of these. Hence the remarkable number and great talent of executive engineers, by whose exertions a vast net-work of irrigation channels was spread over the face of the entire country. All this has operated powerfully in producing the social prosperity for which the irrigated districts are remarkable. In Spain and the south of France, and considerably in Belgium, irrigation is extensively practiced, so that it may be said that the great valleys of the Po, Adige, Tagus, and Douro are subjected to systematic irrigation, enormously adding to their productiveness. Such a system is entirely impossible where the right of the land-owner on a stream to own and control the water is admitted. The water is conducted for miles away from the stream, and from the land of the riparian proprietor. He may have his share on the terms of other users of the vital fluid; but he cannot claim a superior right because his land is nearer, or better situated than another's. And he has no power to determine that the water shall run idly by him to the sea, and lose nothing by nonuser. Such doctrines may do for humid countries, where water is an obstacle; not for arid countries, where it is the supreme blessing—the essential of the community's preservation.

"The climate, productions, and general characteristics of these countries resemble strongly those of California, especially of the southern part of the State. A system that has made possible their dense populations must be favorable, it must be indispensable, to our prosperity. Our population is thin, compared with that of our sister States. We have a cultivable area equal to New England, New York, and Pennsylvania, with a population of a million, while theirs is fourteen millions. Compared to the populations of other countries of the world, which resort to irrigation, ours is insignificant. If we are to observe the law of growth, we must have its conditions. We cannot maintain a population beyond our means to feed. We cannot feed a large population without irrigation, or with irrigation only on narrow ribbons along the river beds, which the Supreme Court permits to riparianists only. Imagination cannot depict the horrors of famine, misery, and death that would follow this rule, sternly applied to the plain of the Indus or of the Ganges. It would produce a revolution if enforced in the basin of the Po. With similar climatic conditions, our present interests and future necessities run parallel with those of other arid countries, not with those of humid regions, like England and the Atlantic States. In the maxims and practice of countries resembling our own in this particular, we may find useful guidance. Our great plains and valleys must be utilized; our foot-hills must be clothed with cultivated verdure; our streams must be taken from their useless and shifting beds and given the widest scope. Then we may create an empire here, of health, prosperity, and development, while the alternative is a dwindled population and wasted resources. The better work had made good progress before the halt called by this decision. It may not be doubted that it will be resumed, and any obstacle will be legally swept away by imperious public necessity, like chaff from the threshing floor."

CHAPTER XXIX.
BIOGRAPHIES.

IT has been said that the reading of "Plutarch's Lives" effected the social and political revolution of France. If the proper study of mankind is man, it is a study to which the biography of every one who has gained fame or fortune is the contribution of a text-book. In the New Empire are scores of men who have attained both, by their genius and their industry. Many of them were pioneers to the Pacific Coast, and others were early occupants of new mines, or the first to perceive the promise of investments which others had passed by, and so in one way or another, and notably by ways honorable and upright, these men have reaped the rewards which crown the genius of industry.

Nations are made up of individuals, and nations that are ruled by constitutional forms have governments that are the result of accepting successful experiments as their model. If all exploits in the science of government that were proved to be failures, had been accepted as precedents to be copied and imitated, statecraft would have been an aggregation of mistakes, a hump-backed and reel-footed science, and human government would be now a case of chronic rickets, instead of a system growing yearly to a more refined adjustment to the manifold necessities and useful diversities of the race.

Applying this use of the example of success to the individual, there is a well-defined utility in studying the lives of successful men, and in each of such lives there must be some noble elements which teach and exhort. The study of "Plutarch's Lives" was a stimulus to the intellect of France, not because it made of any Frenchman an Alexander or a Cæsar, a Cicero or a Publicola, but because it moved Frenchmen to make the best use of the opportunities within their field of action. So the few examples for which we have space here, it is believed, will move the growing generation on this coast to sustained effort, to hope under adverse circumstances, to courage in the face of difficulties, to the endurance of defeat with patience, and to the celebration of success with moderation.

We deal, in this list of worthies, with four men who have reached and now occupy seats in the Senate of the United States. We select them because of the intrinsic worth of their lives, as examples of the value of readiness, address, and application to the conquest of difficulties which others avoided, and also to correct a prevalent Eastern impression that these gentlemen are in the Senate only because they are rich. They are there because they represent the business classes of the New Empire. In them are the qualities which have conquered, and will continue to win, successes in our commercial and financial activities. In the East the lawyer is universally the recipient of public honors. The successful lawyer is selected. The presence of lawyers in the Senate has made it necessary to propose a law that they shall not practice in cases which may be affected by legislation. Surely it is to the credit of the New Empire that it puts its business men into the Senate, and that they have great fortunes proves only that they are great business men. The East is welcome to its senatorial lawyers; we make no issue against the practice of sending them there, but we refuse to admit the propriety of an issue made upon the representatives of that keen and unconquerable business genius which has developed on this coast, within less than forty years, the institutions of a great and refined civilization.

JOHN P. JONES.

Senator Jones is peculiar amongst the coast senators by reason of his long, continuous service, being now in his third term. It is believed that except for the constitutional provision which bars him out, his public career would have carried him to the Presidency. But he was born abroad, in Herefordshire, England, and although his parents brought him to this country the following year, and he is in all things an American, his alien nativity closes against him the two public offices which are higher than the Federal Senate.

His father was a marble cutter, and on landing in this

country in 1830, pushed westward to Cleveland, where he established his trade and reared a family of thirteen children. The future Senator was educated in the public schools of Cleveland, and that he made the best use of their opportunities is proved by the keenness and culture of his intellect. His education was carried on at the same time he was mastering his father's trade; for in those days the owner of a manufacturing business or handicraft was not forbidden to teach it to his sons. After leaving school, he worked for his father, and also got some practical insight into finance by employment in the counting-room of a bank. In 1850 he and his brother Henry decided to come to California, and the way they chose to make the journey sounds like a fable. The bark *Eureka*, which had been in the Lake Erie trade, was fitted out at Cleveland, and with the Jones boys as passengers, went out through the Welland Canal into and down the St Lawrence into the North Atlantic, and then around Cape Horn to San Francisco Bay. They left Ohio in the early spring, but the summer, autumn, and winter of 1850 were long spent before their voyage ended. They went at once to the mines, and on Feather River, and in Yuba, Calaveras, and Tuolumne Counties got that practical experience which later on was to serve the Senator at the turning point of his fortunes. Wherever he was, he was noted for his studious habits. In this respect he greatly resembled Col. E. D. Baker, who in camp and cabin was always using the best means at hand to increase and enrich his store of knowledge. In this trait they both were of intellectual kin to Daniel Webster, whose retentive mind demanded constant additions to its full treasury. Young Jones, though scarcely more than a boy in years, was shortly elected to the magistracy; a little later he was chosen sheriff. Then came trouble with the Indians, and he volunteered, and did some good fighting. He now began to be known as a public speaker of quite unusual power, a well-equipped debater, a lover of fair play, and tolerant of the views of others while able to maintain his own with a vigor that made him a formidable antagonist. His legislative experience began in the California State Senate in 1863, and he served until 1867, when he ran for Lieutenant-Governor, and went down, leading his ticket. This reverse, for the time, arrested his public career and threw him back upon business. In 1868 he was made superintendent of the Crown Point Mine, the oldest of the Gold Hill Comstocks. It was a property that had not paid for years, and its abandonment had been seriously considered. The mine communicated with the Kentuck and Yellow Jacket, and in the first year of Mr. Jones' control, the firing of the Yellow Jacket caused the greatest catastrophe in all the history of mining on the Pacific Coast. The fire was in the 800 feet level. The day shift had nearly all gone below, and forty-five men perished. In this emergency the superintendent showed himself entirely a hero, and to-day throughout Nevada and in every mining camp on the coast, the story of his courage and humanity is told over and over again. It was the foundation and beginning of his popular hold upon Nevada, that has given him longer continuous service in the Federal Senate than any man from the West has ever been permitted to enjoy. After rescuing a great many cage loads of miners, on the second day of the fire, it became necessary to send some one to the bottom of the 800 feet level to cut the air pipes. Mr. Jones went himself, accompanied by a boy who volunteered to hold the candles. Stepping on the cage, they were lowered into that pit of smoke and flame. In twenty minutes the return signal was given, and they reached the top barely alive.

He had faith in deep mining, and pushed Crown Point down to 1,300 feet, where, lying in a solid body 200 feet long, he struck the first bonanza, and that instant became a millionaire. This bonanza yielded $30,000,000.

Mr. Jones was active in developing Nevada properties. His ore mills soon yielded him an income of $30,000 a month, and his money was not sequestered, but was put into productive enterprises that employed labor and stimulated commerce. He did not get money to play the miser, but his generosity increased with his riches, against the rule which yokes wealth and parsimony too often together. In 1873 he was first elected to the Federal Senate, and in that body his worth was soon recognized, and his influence on more than one occasion has determined the course of important legislation. Had it not been for him, the restoration of the silver dollar to our coinage would not have been accomplished. He opened that great discussion in a speech delivered April 24, 1876, and he closed it February 14, 1878, in a speech that will live as long as the precious metals preserve their universal desirability and are sought by man. In this speech were many gems, but his vindication of the miners of this coast will be cherished the world over as the miner's best certificate of character. Senator Jones opened this debate to one Senate, he closed it to another; but in the two years' debate that lay between not a single salient point escaped him, nor did he lose sight of a single feature in the procession of events throughout the world which during that time would illustrate or enrich his argument. The position he gained in that discussion fortified him in the respect of his supporters and his antagonists. His frankness of nature revolts against the lines of party when they seem to him limits to truth. His declaration on the race question was the keenest analysis of the relations of whites and blacks in the South that has ever been made, though it was against a tenet of his party.

His official duties have never been neglected, though his private affairs have drawn upon his energies. Helpfulness to friends, and a desire to stimulate the industrial activities of the coast, have somewhat impaired the wealth won by his boldness as a practical miner; but patience and prudence and a courage unfaltering, have relaid the foundations of a fortune for him that promises to be the largest yet amassed in the New Empire.

The Senator's wife is worthy to share his honors and his fortune. His own social attractions are very endearing, and Mrs. Jones, accomplished and charming, presents the quite uncommon spectacle of a brilliant as well as amiable wife to a brilliant and amiable man, and this must be the reason why to their friends their life seems a courtship and their home graced with contentment.

GEORGE HEARST.

Senator Hearst is a fine example of clear grit. There are some men whom fortune downs, and they stay down. There are others whom the fickle jade may trip, but they will not stay tripped. She has tried more than once, in finance and politics, a catch-as-catch-can with George Hearst, but he never stayed thrown, and now is beyond risk of misfortune in any encounter of that kind. In descent he is of parallel lineage with Andrew Jackson and John C. Calhoun; for his and their Scotch ancestors settled together in South Carolina at about the same time, and their careers ran together. Mr. Hearst's father was a native of South Carolina, who settled in Missouri while it was a Territory, and there George was born September 3, 1820, making him eleven months and eighteen days older than the State of Missouri. He got the sturdy experiences and hardy lessons of frontier life, and made good use of the primitive school facilities of the new country. In 1850 he came to California overland, made money in placer mines, and went broke in quartz. Getting a stake in the placers, he went back to quartz, and was mining in Nevada County when the Washoe excitement broke out in 1859. He had seen the black ore from Mt. Davidson, and an assay proved to him that it had in it silver at the rate of a dollar a pound. He got an outfit, and crossed the ridge into what was then Utah. He found only a score of men on the ground, and the prospecting had gone as far as only a few pits in the ground, not more than four feet deep. Hearst was almost the only one who knew the value of the ore for silver. They were all after gold. He remained six weeks, and decided that the discovery was of immense importance, took an interest in the Ophir, and went back to Nevada City to get the money to pay for it. Returning, he began work on his claim, getting out the free gold with a Mexican *arrastra*, and sacking the pulp for shipment to San Francisco. After sending down forty-five tons at a freight cost of $500 a ton, they found it could not be sold at any price. This reads like a fable at this end of the output of the bonanzas, but it is the sober truth. At last a bold metallurgist agreed to work it for $450 a ton. It yielded $3,800 a ton, and when this was coined into silver dollars at the mint, it settled the destiny of the Washoe country. Mr. Hearst sold half of his claim and bought more, and in 1860 revisited Missouri to support the declining years of his mother, and during this dutiful sojourn married the very accomplished lady who has since cheered and greatly guided his career. Returning to California in 1862, he resumed mining on the Comstock, and by 1865 was a millionaire. In 1866 he was for the third time downed by financial reverses, but going into San Francisco real estate, he soon got ahead a few hundred thousand, and went back to mining, in which he has since made a matter of $20,000,000. He owns mines or mining interests from Dakota to Mexico, and his income is put at $2,000 a day. His public career began in 1865 as a member of the California Senate. In 1882 he was a candidate for the gubernatorial nomination, but was defeated by General Stoneman, who in turn appointed him to the United States Senate to fill the vacancy caused by the death of Gen. John F. Miller. He was in the City of Mexico when informed of his appointment, and went directly from the capital of one Republic to that of the other. His service in the Senate has made him popular in that body and strengthened him at home, so that he is already prominently spoken of in the East as a candidate for the vice-presidency. His friends are confident that if he wants it he will get it, for that is Uncle George's way.

LELAND STANFORD.

Senator Leland Stanford is of English and New English ancestry. His family was on this continent as early as 1644. His father, Josiah Stanford, was a native of Massachusetts; but when he was four years old the march westward began, and the family halted in New York, which was then frontier, and there Josiah grew to manhood, and was a successful farmer. Leland was born on his father's farm, March 9, 1824, and in that morally and physically excellent rural life his youth was passed. He was well educated and approached manhood in that even balance of wholesome mind and person which certify to the fidelity of parents and the tractability of children. In 1845 he chose the legal profession and began its study with Wheaton, Doolittle & Hadley, in Albany. It was to be his destiny not to practice law, and perhaps it is because he did not that his country has in him an intellect as broad as it is vigorous; for it has been well said of law practice that, while it sharpens, it also narrows the mind. However, his legal study and admission to the bar was the perhaps unconscious preliminary survey of a path since held to be necessary to the feet of successful business men; for our best educators defend law schools maintained by the State upon the distinct ground that a law course is the most important part of a business man's education, and that inasmuch as the business men of the country are those whose activities generate its revenues, hold it equal in rivalry with other nations, promote its schools, sustain the different establishments of religion, foster art and equip science for conquest, therefore the State, in its scheme of public education, should consider the best means of their complete preparation for a career which affects interests so thoroughly compacted with the national life. When this argument needs an illustration, it may be found in Leland Stanford.

After admission to the bar, he set his face westward, as his grandsire had done before him, and in 1848 settled in his profession in Port Washington, Wisconsin. Two years later he returned to Albany and married Miss Jane Lathrop. This happy and well-assorted union seems to have put him in the path of destiny. He found the practice of law so different from the elevating study of its principles, and so felt within him capacities for a different career, that he gave up for good a profession which forced him into the disputes of others, and in 1852 furnished that evidence of fitness for great things which every man displayed who came to this coast, remote and little known, under novel conditions of life and commerce, to seek a fortune.

Here he went at once into the trying labors of the State's great industry, and at Michigan Bluff, on the American River, for four years took manfully his share of the toils and hardships of a mining camp. In 1856, with the avails of his mining and in association with his brothers, he began merchandising at Sacramento, and there laid the real foundation of a career which has attracted and charmed the attention of the whole business world. In 1857 he was a candidate for State Treasurer, but was defeated by Hon. Thomas Finley, of El Dorado.

In 1859 he ran for Governor, but was again defeated. In 1861 he ran for the same office again; was elected by a large majority, running 6,000 votes ahead of his ticket, and served with distinguished credit in a time that tried the intellectual and tactical resources to a degree that broke weak men down. This service for the time closed his political career. Incident to the stirring events of 1861, when for a time California had seemed to hesitate in deciding whether her allegiance lay with the old Union, into which her path had been hewn by patriots, or with the new Confederacy, which genius and ambition had just baptized with independence, and committed to the arbitrament of battle, there had developed the need of a closer contact between this State and the East. The military operations of the Government required it, and the time had gone by when the whole region subject to national jurisdiction, lying between the Rocky Mountains and the sea, could be subjected to the means of communication furnished by the control of San Francisco Bay. In this necessity for a more perfect union was the germ of the transcontinental railway. Of this wonderful achievement of human energy and genius and courage, we have elsewhere treated. Let it be said here that without the calm and inflexible spirit of Leland Stanford, the Sacramento merchant, no part of the transcontinental system of railways would have been built or controlled by California capital. But for him this national convenience and coast necessity would have been created and owned by Boston or New York, to serve as a siphon that should drain our profits and avails to the East, and make no return. The story is too long to tell. In the beginning Governor Stanford and his associates were sneered at, guyed, and traduced as visionaries. The Sierra Nevadas were believed to be impenetrable and impassable, and if they were passed, beyond lay the weary, dreary desert, which the Forty-niner remembered with aversion as the scene of his sufferings and perhaps the grave of his companions. Nearly everybody said that it was against common sense, this attempt to build such a railroad, for did not the snow sometimes lie on the Sierras forty feet deep! Did not the Donner party die, or live in worse than death, right where this line would run! But those things disapproved or undiscerned by common sense are favored and clearly seen by the keener vision of that sense which is not common. And so through appalling difficulties, financial and topographical, the road was pressed to a finish. Wearied by the burden, the story goes that the completed enterprise was put on the market by its authors and finishers, but it found no buyers, nor did its stocks when men in San Francisco were implored to take them. So that which he had builded Governor Stanford was compelled to hold and administer, and in later years many a jealous man has gnawed his heart at the success in which he was offered, but spurned, participation.

Through this successfully managed enterprise, great wealth has come, and Leland Stanford ranks foremost amongst the world's capitalists. In his different operations on the coast he employs and pays wages to between twelve and fifteen thousand men. His wealth has gone into all kinds of constructive and productive enterprises. If he saw a manufacturing or other business languishing for lack of energy or capital, he bought it without haggling, equipped it for success, and made it succeed.

In 1883 he visited Europe, and there, in old Florence, came the unspeakable sorrow of his life, in the death of his son and only child, Leland, a youth of parts most promising, in whom were centered hopes the loftiest and affections the tenderest.

Throughout their life together Mr. and Mrs. Stanford have been known for wisely generous support of the good works of charity and education. In them pity's sweet fountain never ran dry, and its affluence took substantial forms. The kindergarten system of San Francisco, a rich benefaction, grew up under Mrs. Stanford's wise endowments. In their retreats and asylums hundreds of orphaned children have blessed the spirit of motherhood incarnate in her. At her old home in Albany she is building and endowing a home for aged women, at a cost of hundreds of thousands.

On their return from Europe they perfected together a plan long entertained for the endowment of a university.*

In 1885, the California Legislature elected Governor Stanford to the Federal Senate. It was done as a voluntary recognition of his benefactions to the State, his knowledge of its needs, his interest in its development, and his primacy as a business man. The East has talked about our rich Senators. She has men of great wealth in that body, and can it be said of them as of Senator Stanford, that this honor came without the indication of a wish to add it to the laurels of a busy and beneficent life?

In 1884 his nomination to the presidency was mooted, and since his senatorial service has shown the profundity of his experience, his ripe learning, his judicial temper and his executive force, the proposition is renewed to give the country the benefit of his qualities in that great office which is now so ably filled that the succession next chosen must be of superior merit.

The mere politicians will not agree to such a selection. They will conjure objections as countless as the phantasies that come in the dreams of drunkenness or surfeit. But the people may conclude that the man whose genius has wrought out business enterprises which adorn and dignify the century, and whose benevolence has spanned the continent in quest of God's poor, forgotten by the priest and the Levite, and whose culture and conception of its need in others has prompted the gift of tens of millions to found what America has not, a complete university, may also as president represent the refinements of our civilization and the energies of our people. He represents, too, the right use of wealth, which, if gathered in unwise ways and spent in ostentation, affects the masses to a sinister temper; but gained by him in adventure and by making no man poorer, for it was added by his creative energies to the commonwealth before it became his personal possession, and spent in the spirit of a Christian stewardship, his wealth was won in ways that benefited others and is devoted to the good of mankind.

---

* We cannot more completely describe the extent and intent of this university endowment, than by reproducing this editorial from a San Francisco paper, printed a few days after the gift was passed to the Trustees, in November, 1885.

"On the 14th instant this city was the scene of an educational foundation that is destined to be the initiative of the most richly endowed institution of learning in the world.

"There has been talk of some public recognition of this benefaction that shall take the form of a permanent memorial to the two founders of Leland Stanford, Junior, University; but it occurs to us that their memorial is already provided in the institution itself. John Harvard, two hundred and forty-seven years ago, gave $3,500 to the college that bears his name, and by that gift purchased an immortality that no monument of granite nor tablet of brass could have preserved to him. So, a hundred and seventy-one years ago, Elihu Yale, by a gift of $2,500 to endow a college, perpetuated his name to generations yet unborn, while the world has already forgotten him as ruler of Madras and Governor of the great East India Company, out of whose monopoly of trade emerged a new empire for England.

"The Stanford University begins its career with greater secured and permanent financial resources than are possessed by any of the established universities or colleges of this country, and as accretions of the capital must continually outrun demands upon it, its treasury will soon be the richest in the world.

"Standing at the hither of this event, its farther consequences are not plainly seen. The prospect is bewildering in its possibilities; for so many growths, so many institutions, and such a variety of virtuous and profitable activities impinge upon that which the great capital is to conjure into form, that the mind is embarrassed in the effort to reduce it to a generalization.

"The university, if it realize its mission, will be not merely the resort of those who seek instruction in letters, the arts, and the technical knowledge which takes within its sweep physics and the manual occupations. It will also be the seat of original investigation, and this, President McCosh, of Princeton, declares to be one of the distinguishing characteristics which marks the university and sets it apart from schools of lesser grade. The university must not only transmit the gathered wisdom of the ages, but it must add to the store. Hence, in this junior of the world's universities, most remote from the center of the Anglo-Saxon race, in the youngest of the great cities of the continent, but blessed above all its fellows in the generosity that lavishes its endowment, we may expect a wonderful impulse to be given to original research in philology, philosophy, physics, and throughout the circle of arts and sciences. Its scheme is precisely adapted to the line along which the distinctive intellect of this side of the continent is developing; for here the artistic sense is as indigenous as in Southern Europe, and the genius of practical work is as defined as in New England. The latter toils patiently to provide the condition of society in which the former may display its results, and it is the high purpose of this endowment to cheer and encourage each. It may be said that such a vast institution must not expect to serve only the population nearest to it, and hence it should be different in some features of its scheme. The answer to this is, that no curriculum can offer a more symmetrical culture than one in which fancy and fact are so combined. The purpose of all labor is the production of wealth, and technical instruction is intended to multiply the working and earning, and hence the wealth-making, power of human hands. But the whole process would be robbed of half its motive if æsthetic culture did not point out the refinements to which that wealth may be applied.

"The universities of France and Germany were adapted, primarily, to their more immediate contacts; and they have continued in that state, at one with the genius of the people in the midst of whom they have withstood the vicissitudes of centuries. It is because they have reflected the best thought of these tributary people, and by the fruits of original research have led to greater growth and trained to constant absorption the intellect that is subjected to their influence, that they are sought by students from all over the world. To provide here for culture in art, letters, and polytechny is to breathe into the ribs of this project the atmosphere that must sustain its growth, and the same results may fairly be expected here that have elsewhere followed like efforts.

"If judged only by the buildings and laboratories and workshops and professional chairs, the scholastic plant, and the number of students it is to nurse into knowledge, or, if measured by the hard problems that shall yield their long-secreted solution to the patience of its original investigators. the Stanford University is seen on one side only, and on that imperfectly, for the present perspective is insufficient. There are certain practical effects which will reach innumerable masses of men who will never see its class rooms nor stand in the shade of its walls, who may, indeed, live and die in ignorance of its existence, while they are its direct beneficiaries. Commerce follows intellectual culture and loves to breathe the same air. When the Italian universities were eminent, commerce sought that land. The East, the cradle of the human race and the source of wealth easiest won, saw the Attic beacon on the Calabrian Peninsula, as the Magi saw the star of Bethlehem, and gave its spoil to freight the argosies that made Venice the proudest commercial city of her day. When trade abandoned the Adriatic, leaving behind colossal fortunes that are not yet exhausted, and monuments of architectural taste and fairy interiors that are yet unmatched, it was loth to leave learned Italy. True, in its wake had been Shylocks and Antonios, but there were also the learned doctors of Padua and Parma. Passing into the Mediterranean, commerce furled her sails at Genoa, and rested there so long that both sides of Italy had been gilt with the profits that followed the excellence of her schools before it took flight and passed the pillars of Hercules, to thrive in the superior luster of Leyden and Utrecht. As at Venice culture and commerce joined hands in building a city on the bosom of the sea, they were in Holland partners around the Zuyder Zee in creating wealth which advanced dyke and dam against the ocean, and reclaimed from the sea whole provinces of land, and built cities where fleets had floated.

"Finally, Oxford and Cambridge drew commerce to the Thames, and made London the world's commercial capital.

"The trade of coasts and continents follows the same law, and on this Pacific side of the two Americas has waited for some such supreme manifestations as this foundation to be attracted to a common center.

"So, when Senator and Mrs. Stanford, acting upon a long-formed plan, at last perfected in a noble sorrow, gave millions in trust for higher culture, they were not only building the walls of a university and endowing its chairs, but the pens that signed away a great fortune were keys actuating a web of wires unseen, running to myriad consequences that were not named in the passage of this mighty gift. In that act they were turning the glebe, they were planting virgin acres with seed, and were enlarging upon this round globe the gilding of the harvest. They were opening new mines, and were stripping fresh quarries to flux noble ores. They were heating the cupola and giving impulse to the currents of molten iron and steel that flow into cunning moulds, to be shaped to many a profitable purpose. They were inspiring with motive the brawny arm that makes the anvil ring, and rewarding the cunning hand that shoves the plane and guides the chisel. They were rousing the shipyard's activity and preparing the launch that weds to the water many a stately ship. They were throwing the shuttle of countless looms through warp and woof of cotton and of wool, and giving distant shepherds dreams of plenty, and cheering with right reward the dark-skinned toiler between the snowy rows of cotton. They were planting vine and olive, and corn and wine and oil will join in sacramental sanction of an act that in its incidents shall build many a home, with fire on its hearth and bloom on its lintel and its threshold pressed by happy feet.

"Their own generation may not have a full conception of the import of what they did as it is shut out from witnessing or sharing all the crowding consequences that are to come; but in their act is latent the luxury of thousands, the comfort of coming millions, because natural laws are irrepealable, and cultivated intellect is to-day the founder of States and the promoter of commerce, as it was when Moses led Israel to the promised land because he was learned in all the knowledge of the Egyptians."

JAMES G. FAIR.

The youngest of our four Senators from California and Nevada, is James G. Fair, born in far Clougher, County Tyrone, Ireland, in the last month of 1831. So vigorous and alert is he that he seems hardly to have yet passed his youth

He came to America young, and was located in Geneva, Illinois. When Chicago began to be regarded as a business place, to that embryo city he resorted to get into business. He did gain there experience and training which filled him with aspirations which they also fitted him to attain, and at the age of seventeen he joined the long procession overland to California. Starting in 1849, the long journey ended in 1850, and he was soon swinging a miner's pick in Plumas County, at Long Bar. He followed his mining instincts from prospect to prospect, and acquired that varied experience in all kinds of mining and all forms of gold and quartz, which was his preparatory school for the grand opportunities of the Comstock. By 1860 he reached Virginia City, with some money, more experience, and the most faith of any man who had early contact with those strange and defiant ores. With this equipment he had confidence when others lost it, and he bought the claims of the doubting and the thriftless, believing firmly in the outcome. So he owned interests in, and became superintendent of, the Ophir, and Hale, and Norcross, and later on around the properties he had believed in, and hung to, the Bonanza firm was formed and he was a partner. In his Hale and Norcross was made the first half million of the multiplied millions which that firm took out of the Comstocks. Under Mr. Fair's advice, more claims were now acquired to consolidate and extend their properties. Then this shrewd miner, drawing upon his knowledge and expert faculty and upon the faith of his partners in him, began that profound search which uncovered the first bonanza and gave the firm one hundred millions! This partnership is now dissolved. It stands alone in ancient history and modern in the magnitude of its operations, the absolutely fabulous wealth it found in minerals, and in the private fortune which fell to each of its individual members. Senator Fair's capital has gone into eligible real estate, and since he became a member of the Senate he has quietly pushed into new fields of activity, which promise to yield greater results, even, than he gained in the Bonanzas. He has one by one bought all the interests in the South Pacific Coast Railroad, which is already the most extensive narrow gauge system on the continent. It has adequate terminal facilities on San Francisco Bay, and its passage through Oakland has stimulated improvement in that city to an extent unknown for years. Cable roads collateral to it are being built, and the population of Oakland is getting large accessions in the prospect of that prosperity which many railroads bring to a commercial center. But

Senator Fair is not building a railroad merely to enhance the interests of Oakland. His own State, Nevada, is the ultimate beneficiary of his enterprise, and when he has given her a narrow gauge line to San Francisco, and furnished the trunk line of a system so well adapted to the penetration of her valleys and the scaling of her mountains, Nevada will be covered with a network of narrow gauge feeders to this main line, which will help the development of her mines and encourage her growing agriculture. Such a road and its collaterals are the present vital need of that State. The East never tires of girding at Nevada. It is denounced as a "rotten borough," and Rhode Island and Delaware turn up their little noses at it. Its disestablishment has been agitated, and its brave and hardy people have been taunted with decreasing population and receding prosperity. The best friend Nevada can have just now is the capitalist, who will put all the resources of her soil, grazing, glebe, and mineral, within reach of those who want to develop them, by just such a transportation system as Senator Fair has in hand. But let no man fancy that this narrow gauge road will stop in Nevada. A narrow gauge line is building from the Missouri River toward Denver. From Denver to Salt Lake is already in operation the Denver and Rio Grande narrow gauge, and grading for a narrow gauge from Salt Lake already reaches far westward. So it is manifest destiny that Senator Fair should be at the head of a transcontinental narrow gauge railroad, than which no business venture can be of greater interest to the commercial community, while its saving grace to Nevada is indisputable.

In the Senate Mr. Fair has influence ranking with men of the first class. Suave and thoughtful of others, he is a great social favorite, while the Senate consults his views upon a business question and adopts them with confidence. Affable, approachable, and zealous, the New Empire counts him amongst the major forces in commerce and public life, upon which she relies to push forward the frontiers of prosperity.

FINALE.

We have now passed in review the characteristic features of the New Empire. In area it is so large that all of Europe would be hidden in it. Its mountains are noble in their aspect, and are necessary and useful features in its climatolgy. Its railway system is being rapidly approximated to the full measure of present needs, with facilities for ready extension to accommodate the future, while its river traffic, its coast and trans-Pacific steamers, its noble argosies of sailing craft, complete a system for the convenience of travel and traffic that is unsurpassed.

Of all this, and all that is to be, San Francisco is the commercial center. Here the Central and South Americas, the Islands of the Sea, and Asia, will bring their trade, and the Golden Gate will receive the commerce of the world.

Here a high civilization will always have its seat, and there is room for millions of people to plant and sustain its institutions. The fact will be demonstrated that the Pacific side of this continent can sustain a denser population than the Atlantic Coast, or the interior, because our Asiatic climate and fertility imply an Asiatic approximation in the density of population. The East is the analogy of Northern Europe, and resembles it in the characteristics of its people.

If the first settlement had been on this side of the continent, the East would be now far less known, because far less desirable than is the New Empire.

We have here great men and excellent, cultured women, and the material resources which call out the talents of State makers and home builders.

We have outlined the advantages which here await the settler and the inventor, and have aimed to wisely guide the inquiries we hope to have stimulated. We have as frankly pointed out the obstructions to the future as the remedies which we believe will overcome them, and have dealt in thorough candor with our readers. But, after all, the half has not been told, and the New Empire has to be seen to be appreciated in all its merits.

If we have persuaded our countrymen that Nature here spreads beauties of scenery which should be enjoyed in preference to the lesser graces of Europe, and if we have convinced any of the comforts and pleasures of life in this winterless land, our work has in accomplishment fulfilled the benevolence of its intention.

www.ingramcontent.com/pod-product-compliance
Lightning Source LLC
Chambersburg PA
CBHW020247170426
43202CB00008B/255